Advanced Networks, Algorithms and Modeling for Earthquake Prediction

RIVER PUBLISHERS SERIES IN COMMUNICATIONS

Volume 12

This series focuses on communications science and technology. This includes the theory and use of systems involving all terminals, computers, and information processors; wired and wireless networks; and network layouts, procontentsols, architectures, and implementations.

Furthermore, developments toward new market demands in systems, products, and technologies such as personal communications services, multimedia systems, enterprise networks, and optical communications systems.

- Wireless Communications
- Networks
- Security
- Antennas & Propagation
- Microwaves
- Software Defined Radio

For a list of other books in this series, see final page.

Advanced Networks, Algorithms and Modeling for Earthquake Prediction

Massimo Buscema
Marina Ruggieri

River Publishers
Aalborg

Published, sold and distributed by:
River Publishers
PO box 1657
Algade 42
9000 Aalborg
Denmark
Tel.: +4536953197

ISBN: 978-87-92329-57-8

Dedication

"To Angela, Antonio, Danilo, Fabrizio, Maurizio, Remo, Romeo and all the other friends that experienced the terrible Earthquake in L'Aquila on April 6, 2009."

Marina

"... and to all the persons who feel that quakes are predictable, but are not able to show how"

Massimo

"Science cannot be used against the human life and dignity, but must be used at the service of future generations. Scientists have ethical responsibilities and the conclusions of their studies must be guided by the respect for the truth, by the honest recognition of the accuracy, and the inevitable limitations of the scientific method. This means avoiding unnecessary alarmist predictions when not supported by sufficient data or in situations exceeding the current forecast capacity of the science; but also means avoiding the opposite silent for fear of facing the real issues: the influence of scientists in shaping the public opinion on the basis of their knowledge is too important to be undermined by improper haste or from pursuing superficial publicity." (Benedict XVI, 06/11/2006)

Editors Biography

Massimo Buscema (1955) Professor and Computer Scientist, expert in Neural Networks and Artificial Adaptive Systems. Director and Professor at Semeion, Research Center of Sciences of Communication, Rome (Italy).
(from 2010-today) Member of the Advisory Board of the Center for Computational and Mathematical Biology (CCMB) at University of Denver, Colorado, USA.

(2009–today)	Consultant of the italian "Presidenza del Consiglio dei Ministri", Rome, Italy.
(2003–2007)	Consultant of New Scotland Yard, London, UK.
(2003)	Nominated "Grande Ufficiale al merito della Repubblica Italiana" by the President of Italian Republic.
(2000–today)	Consultant of Bracco Pharmaceutic Group, Milan, Italy.

Member on the Editorial Board of various international journals. He has designed, constructed developed new models and algorithms of Artificial Intelligence. Author of scientific publications on theoretical aspects of Natural Computation, with over 250 titles: scientific articles, essays, and books on the same subject.

Inventor of 20 international patents.

Marina Ruggieri is Full Professor in Telecommunications at the University of Roma Tor Vergata, where she also directs a M.Sc in Advanced Satellite Communications and Navigation Systems. She is Director of CTIF_Italy, the Italian branch of the Center for Teleinfrastruktur (CTIF) international research network. She is President of the IEEE Aerospace and Electronic Systems Society.

Her research focuses on space communications and navigation systems, integrated systems, mobile and multimedia networks, ICT for biotechnology, energy and disaster ahead management.

Since March 2010 she is member of the Committee of Experts for the Research Policy (CEPR) of the Ministry of University and Research (MIUR).

She is author of about 300 papers, on international journals/transactions and proceedings of international conferences, book chapters and books (8).

Preface

When professors Marina Ruggieri, University of Tor Vergata, and Massimo Buscema, Semeion Research Centre, both located in Rome, asked me to be associated with them in drafting a book dealing with the study of the precursors of seismic events, I felt a bit worried as I am an engineer and I had acquired only the knowledge of geology necessary and sufficient to carry out my duties during the course of my career in the Armed Forces. Nevertheless, at the end and after my initial hesitation, I accepted the challenge, considering that I was born in Naples and in an area called "Campi Flegrei", well known for its volcanic nature, and that I was educated in Pozzuoli, near the famous "Solfatara Puteolana".

The "Solfatara" is one of forty volcanoes that make up the "Campi Flegrei" and is located about three kilometers from the center of the city of Pozzuoli, not far from Naples. It is an ancient volcanic crater, called "caldera", still active but dormant since nearly two millennia; it retains an activity of fumaroles of sulfur dioxide, jets of boiling mud and is characterized by a high soil temperature. Similar activities are also present in many other locations in the world: by antonomasia, they are indicated by the name of "my" Solfatara, for their resemblance to the Puteolana one.

The Solfatara is an outlet valve for the magma under the Campi Flegrei area, a valve which allows a constant pressure of gas to be kept in the underground. This area is very famous for bradyseismic uplift and subsidence. The inflation and deflation of the Solfatara is especially well documented due to its seaside location and a long history of habitation and construction in the area.

In particular, in the city of Pozzuoli (precisely in the Roman Temple dedicated to Serapis) there are three marble columns that have boreholes made by marine mussels: these boreholes can be seen on the surface of the columns up to seven meters, showing how bradyseism in the area lowered the ground to at least this depth under the sea level and subsequently raised it again.

In their dialect, local people call this phenomenon "dry tides". More recently, from 1968 to 1972, the town suffered an episode of positive bradyseism and rose by 1.7 meters. Another rise of 1.8 meters occurred in the years from 1982 to 1984. I witnessed both episodes.

Several times, visiting the Solfatara or the roman remains of Pompei, or walking along the seafront promenade of Naples, grandparents and parents told me of Vesuvius and its plume, which unfortunately I have only seen in photographs.

On the other hand, I still have vivid memories of the seismic events which I witnessed in my youth in the city of Naples, and later, in maturity, of the events in Umbria and Abruzzo that awakened me overnight in my home in Rome. These memories of my youth convinced me to try and offer a contribution to the project of Ruggieri and Buscema by drawing on my various experiences as an engineer, thus continuing the cooperation with the two professors I worked with to contribute to rebuild the laboratory of geomatics at the University of L'Aquila.

The idea that these studies could help the international scientific community to find, one day, the key to predict to some extent the seismic events, seduced me. As a matter of fact, this specific subject is very peculiar and sensitive, especially as in recent times it has caused much controversy in Italy due to some incorrect interpretations or use of the words "forecast" and "prediction".

"Forecast" is the word we are accustomed to and that is currently used in meteorology for weather "prediction". With "forecast" we mean "see first" the evolution of a framework of very complex situations on the basis of similar frameworks that already occurred in the past. The "forecast" that it will rain tomorrow or the sun will shine in a certain area is not clearly a prediction, i.e. it has not an absolute value, but it is only a probabilistic estimate of weather behaviour. In weather forecast, through both empirical and historical knowledge, thanks to the application of increasingly complex mathematical models, the professional meteorologists can reach the conclusion that in a certain area and in a certain time interval there is (for example) 80% chance of rain. Weather forecasts have proved to be increasingly more accurate in both the short and the long term (one week) periods and are constantly improving mainly thanks to the observations and data from meteorological satellites.

Nevertheless, these are weather forecasts and not weather predictions. No serious meteorologist will ever indicate a definite amount of rain that will fall in a given location at a precise time. In some cases, some attempts of weather forecast are made on the occasion of Formula 1 races, trying to predict where and when the rain would fall, with predictions based on observations of just a few cubic kilometres of atmosphere over areas of few kilometres, if not hundreds of square meters. For the sake of motorsport, even these forecasts are still quite unreliable.

Meteorologists, however, have a great advantage, as they deal with a system, the atmosphere, which is very fluid and directly accessible and measurable by means of a number of satellite sensors and weather balloons, with a lot of historical data accumulated over years of soundings and observations. Furthermore, for sixty years they have enjoyed the benefits of the World Meteorological Organization, under the auspices of the United Nations.

This semantic explanation is necessary to underline that the current state of the art can not even allow predictions about changes in the weather and, therefore, a fortiori, in a near future it will not be possible to make predictions about earthquakes. Having said that, it is also my personal belief that just as we can now make forecasts for weather, maybe one day we will be able to make forecasts for earthquakes. The basic idea is to apply to the Earth a methodology similar to the one applied to the atmosphere, that is collecting data from various sources, including Earth observations from satellites, developing algorithms and mathematical models suitable for data fusion, then modelling the evolution of the slow-moving Earth surface considering it as a high density fluid, in order to evaluate "possibilities of an earthquake occurring in a given area in a certain period of time".

The study of long-term probability is already a major activity of institutions such as the National Institute of Geophysics and Volcanology in Italy, as well as many other institutes in other countries with great seismic culture, like China and Japan. The aim of this book is to analyze the possibility of a short-term probabilistic study.

In this preface some of the methods and the theories currently known to monitor the seismic precursors, even if not universally accepted, are listed:

- monitoring the chemical parameters of soil and groundwater;
- monitoring the subsurface electrical conductivity and polarization;

— monitoring the seismic waves;
— applying statistical methods (for the medium to long term), based on analysis of historical and spatiotemporal data;
— monitoring the deformation of the Earth crust;
— observing the electromagnetic waves at Extremely Low Frequency (ELF);
— radiogeophony;
— acquiring knowledge of the geological fault structures;
— applying prediction models on the changes induced on the Earth crust by the gravitational influence of the Sun and the Moon;
— observing anomalies in the behaviour of animals;
— observing morphological and physical-chemical changes of plants;
— observing the Earth's geomagnetic field;
— observing the ionization on the ground and in the upper atmosphere.

Professor Marina Ruggieri and Professor Massimo Buscema, editors of the book, intend to approach the study of earthquakes applying high technology ICT and Artificial Intelligence, probably for the first time in history. They gathered different professional people, geologists, mathematicians, engineers, and officers of the Italian Defence to build up an interdisciplinary team to say something new in the timeline of geology and volcanology.

The techniques of Earth Observation (EO) with Synthetic Aperture Radar (SAR) satellites in low orbit (COSMO-SkyMed) and the possible use of SAR images for the detection of surface movements of the Earth crust due to pressure of tectonic plates in slow motion are recent conquests.

Tectonic plates are able to move because the Earth's lithosphere has a higher strength and lower density than the underlying asthenosphere. Lateral density variations in the mantle result in convection. Their movement is thought to be driven by a combination of the motion of seafloor away from the spreading ridge (due to variations in topography and density of the crust that result in differences in gravitational forces) and drag, downward suction, at the subduction zones. A different explanation lies in different forces generated by the rotation of the Globe and tidal forces of the Sun and the Moon. The relative importance of each of these factors is presently unclear. As Earth Observation gave an important contribution to the study of the atmosphere to improve weather forecast, we hope that satellite observation can help finding the contribution of some or all factors causing the drift of the tectonic plates. And now we hope you will find this book interesting.

Enjoy your reading.

Pietro Finocchio

Contents

Part I

Approaching Earthquakes with ICT and Artificial Intelligence

1

Introduction

Massimo Buscema* and Marina Ruggieri†

Semeion — Research Center of Sciences of Communication, Rome, Italy
†*University of Roma Tor Vergata — Center For TeleInFrastructures (CTIF_Italy)*

Earthquakes have always inspired a respectful fear in human beings. Every time they happen, they show dramatically the irreversible power of Nature as opposed to the intric weakness of the human race.

Imagination depicts earthquakes as a mysterious and magic matter. However, as scientists and technical experts, we do have to consider them also from a different perspective: they are natural phenomena that evolve with time and depend on a number of variables.

Their modeling can help us to reply to the simplest and — at the same time — the most complex question: *are earthquakes predictable*?

In case the answer is affernative, which could be the role of the extremely mature Information and Communication Technology (ICT) in setting up an effective prediction process? How Artificial Intelligence Algorithms can contribute to the picture?

God does not play dice with the Universe, someone said. And the dynamics of the Earth is part of the Universe. What we perceive as a random walk, from the point of view of nature might be a logical and elegant dance. The instrument to understand this dance is the language of mathematics.

Obviously, we do not believe that mathematics is a new God. We are trying to follow the "common sense": if I know the mass and the momentum of a body I can predict everything about it, if this body works in a flat land without borders. But if the same body is placed into a dynamical warped landscape, with local singularities, the situation is completely different: I cannot make prediction about its future, because I do not know the function representing its landscape. But I cannot say: because I am not able to understand, so the dynamics of this body is random. It should be the same if the first "homo sapiens" had said that birds are an illusion of the mind or a game of the bad gods, because he was not able to understand their ability to fly. Everything exists, if it persists, needs to be understood.

Nature has proven to be capable of generating many different types of states. It is not useful counting the number of states that nature can produce. That because the very counting act is itself a new state that nature is producing. But nature works also in a world with four dimensions, then it is likely that nature has a finite number of rules to transform one state into another.

Biologic world is the most complex lab where nature allows us to understand its style of work: small elements interact in a collective, parallel and massive way, generating an increasing amount of information and complexity. Nature works from bottom to up. If we implement models able to work only top-down, we will produce only tautologies.

Natural Computation is a mathematics inspired to nature style: a parallel and local interaction among small elements is able to self-organize into complex behaviours: what we call today artificial adaptive systems, artificial neural networks, evolutionary systems, and swarm intelligence. Along this book we will introduce the reader to this possibility: to use a math inspired to the Nature, to understand what usually seems to be unpredictable.

We will try to show that many times in Science what is considered "noise" is a black box wherein the key of the problem is hidden.

Sometimes we play dice because we do not find convenient to cope with the real possibility of a solution.

The book presents our vision about the above matter.

The book is organized in three parts.

Part 1 frames the possible use of ICT and Artificial Intelligence in dealing with earthquake-related Disaster Ahead Management (DAM). The topics are developed in Chapter 2 and Chapter 3, respectively.

Part 2 presents modeling tools for the earthquake issue and proposes possible ICT tools for supporting the earthquake DAM. The topics are developed in Chapters. 4, 5, 6 and 7.

Chapter 4 deals with quakes prediction using highly non linear systems. Chapters 5, 6 and 7 introduce advanced sensor network architectures for earthquake DAM (Chapter 5), semantic tools for the collection of written data (Chapter 6) and satellite sensors (Chapter 7).

Part 3 presents an experimental network for earthquake DAM. A communications architecture (Chapter 8) and a GNSS based network (Chapter 9) are described.

Finally, conclusions and perspectives are drawn in Chapter 10.

2

A Multi-Element ICT-based Approach for Earthquakes

Marina Ruggieri

University of Roma Tor Vergata — Center For TeleInFrastructures (CTIF_Italy)

2.1 Aim

The Chapter introduces a new concept. It proposes and describes an approach — based on Information and Communication Technology (ICT) advanced tools — to support the population and their goods in seismic regions.

2.2 The *Quality of Life* Framework

In the last decades, Progress is characterized by a fascinating contraction: knowledge is more and more specialized and, thus, an effective advancement requires deep understanding in an extremely narrow direction.

At the same time, however, an innovation, a meaningful discovery is increasingly related to the osmosis of concepts, tools and solutions belonging to different areas of the human knowledge.

The interdisciplinary attitude is becoming the most rewarding approach to bring the Human Being to new frontiers and to improve effectively his/her *Quality of Life* (*QoL*).

The QoL is a key-important concept: it deals with a *human being-centric vision*, that puts the technology as a tool to create a better everyday life and to protect the health of our Planet.

This vision is somehow different from a *technology-driven* approach.

When a new technology is introduced, sometimes the efforts of the technical community are more focused to create an application environment than to frame the discovery in a true beneficial path for the mankind.

There are two meaningful examples of technology-driven approaches.

The first deals with telecommunications systems. The latter world has experienced an extraordinary revolution focused on the introduction of user mobility and, therein, on the development of a planetary success: the GSM (*Global System for Mobile communications*) digital cellular system [1]. The evolution of cellular communications from the GSM generation (second generation, 2G) to the UMTS (*Universal Mobile Telecommunications Systems*) generation (3G) has mainly envisaged a technology-driven process [1]. The major challenge has been, then, to create contents and applications in order to match properly the increasing technological capability of the new generation systems.

The second example is related to *Global Navigation Satellite Systems* (GNSS). The transition from GPS (*Global Positioning System*), another planetary success, to the next generation systems, is again mainly technolgy-driven. This results in a large — and fascinating — effort to create applications able to take advantage of the improved system technology and performance [2].

A technology-driven process can certainly envisage applications that are or may become beneficial for human beings.

Nonetheless, the effective deployment of a QoL framework requires a focused design, that shapes the applications on the QoL goal and not only on the technology availability. As a consequence, the QoL-oriented vision may push towards the deployment of new technologies, if the existing ones makes it difficult or even impossible to meet the goals.

QoL and interdisciplinary attitude are, as already addressed, tightly related concepts. That is perhaps why my thirty years of experience in the engineering world and the over twenty years of research in the field of electronics and telecommunications convinced me that I have to give a contribution to improve the QoL of people.

A citizen benefits from a QoL approach if, for example, he/she lives in a *seismic*-prone country — like Italy — with a clear awareness of the risk but also with trust on the availability of technological early warning tools.

Those tools should be capable of both lowering the life-risk and keeping the practical damages related to earthquakes under control.

In what follows a set of elements that can be useful in a seismic-prone environment are identified and linked to available or under development Information and Communication Technology (ICT) tools.

The goal is to make a pervasive use of ICT in a seismic geography in order to improve the inhabitants' QoL.

2.3 A List of Tools

I am not a geologist and I am afraid it might be too late to start a new profession at this point of my life. Nonetheless, in 2009 I experienced as simple citizen the terrible earthquake that destroyed a beautiful city in Italy, L'Aquila, killed 300 people and shocked to dead an entire country. Some of my best friends have been injured or damaged by that quake and I promised to them and to myself that I would have done something.

So, I started posing a simple question: how the achievements and the technology that my research over the years has contributed to develop could be used to improve the QoL in a seismic region?

As a first step, I have identified a short list of tools that might be helpful to the aim. Obviously, the list is all but complete and I do hope that in next years further experts from my field — ICT — and from other fields will add more entries that can be usefully employed to improve the citiziens' QoL in a seismic area.

The list moves from the idea that, beside an obvious deep knowledge on the Earth system and behaviours, a *Disaster Ahead Management* (*DAM*) capability in the seismic region could take advantage from the availability of:

- Present data
- Past data
- Modeling
- Data fusion
- Data processing
- Communications
- User devices.

In particular the listed items refer to:

Present data: data that depict the present situation in terms of conventional seismic parameters and *environmental figures*. The latter can belong to a set of *technical parameters* or to the so called *folk information*. Technical parameters consists, for instance, in air and ground temperature and humidity; folk data can report, for instance, on behaviour anomalies of animals or plants.

Past data: data that depict the situation in terms of conventional seismic parameters and environmental figures (technical parameters and folk information) along a meaningful past time interval.

Modeling: a suitable analytical description of the phenomena and the risk. This is very important to understand, for instance, if seismic events can be predicted and, in case they are, in which time frame.

Data fusion: useful data are very heterogeneous in nature, time frames and, sometimes, quality. An effective mechanism to merge data, without sacrifying their identity, is a very important step.

Data processing: transforming the input data in useful information for the seismic population is the core of the whole process. Processing is a very broad term that includes, in what follows, the use of artificial intelligence algorithms.

Communications: whatever the architecture for the above data collection, merging and processing, the availability of suitable communication links is mandatory to assure connectivity in a timely manner.

User devices: if QoL is the aim of the process, citizens may be equipped with hardware, software or hybrid devices that communicate with them to provide alarms and derive from them part of the data (technical and folk) to be used in the process.

In Figure 2.1 the additional inputs and tools — with respect to conventional seismic information — that can be utilized in a seismic DAM approach are summarized.

2.4 Supporting Technologies for a Seismic Dam

The items listed in Section 2.3 can be effectively brought into a seismic DAM process due to the advanced development or — in some cases — maturity of ICT.

Modern technologies that can be employed in the seismic DAM are, for instance:

- data collection tools
- advanced network architectures
- intelligent databases

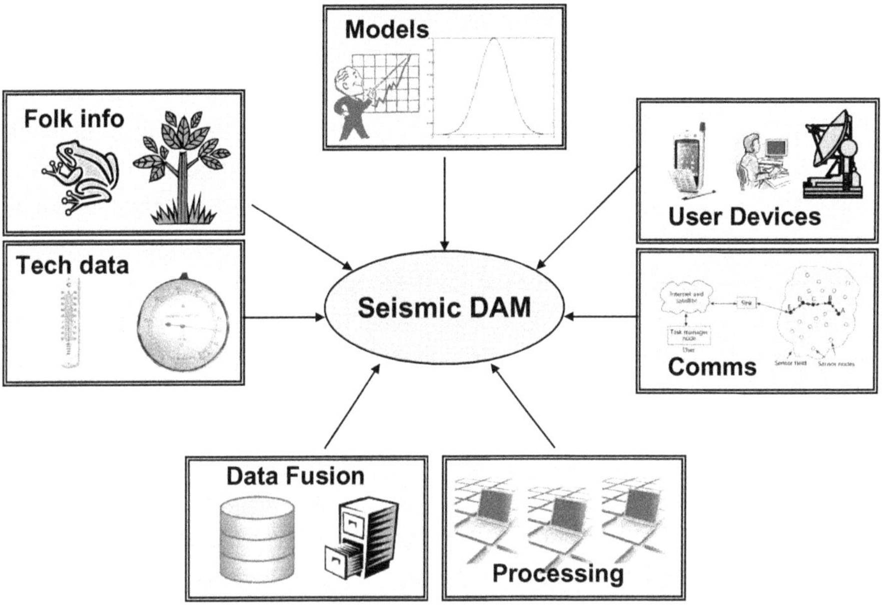

Fig. 2.1 Tools for a seismic DAM.

- adaptive artificial intelligence algorithms
- semanthic analyzers
- integrated communications and navigation user devices.

The above technologies, properly supported by a trustable modeling of the phenomena can be the key to deploy a successful seismic DAM.

I have investigated in more detail those tools and technologies related to the field of interest and expertise of my research center (§Section 2.5)

In parallel, I have interviewed experts of the other fields related to the seismic DAM and I have posed to each of them the same question, in order to investigate if their research achievements over the years could have contributed to improve the QoL in a seismic region.

Surprisingly — not clear yet if I were only very lucky — the experts I have met all agreed about my vision and studied seriously the problem, coming to the analyses and the results that are highlighted and documented in this book.

How the identified technologies can support the listed tools in a DAM approach?

***Present and past environmental parameters*:** the collection of the environmental data (technical and folk data) is supported by the deployment of sensors, linked through a wireless architecture (Wireless Sensor Network, WSN), inside the area of interest. Sensors can be placed on the terrain, inside houses or buildings, on plants, animals and people. Wireless technology is very mature and suitable for the deployment of large, medium and very small networks, like BAN (Body Area Networks), PAN (Personal Area Networks), PN (Personal Networks), as well as their grouping into bigger structures (federations) [3]. The data, that flow through the advanced network architecture implementing the DAM process, are inputted to an Intelligent Database (IB). Folk inputs may undergo a passage through semantic analyzers that translate the information available in written or oral words into a set of numerical data.

***Data Fusion and Processing*:** the core of the seismic DAM process consists of the algorithms and supporting technology that merge the collected data and derive a possible provisional path for the seismic phenomena in the area of interest. Very advanced algorithms are required to this respect, such as the adaptive artificial intelligence systems.

***Communications*:** the network architecture plays a key-role in the implementation of the seismic DAM process. This suggests the use of some among the most advanced and flexible schemes, i.e., a networks based on the Distributed architectures and the Swarm organizational principles (D&S net). The network is crucial for the timely and trustable connections among the major players of the DAM process, including data collection system(s), database(s), adaptive artificial systems, DAM alert center and citizens. The D&S architecture may merge in some of the nodes the communications, data collection (WSN) and positioning functions.

***User devices*:** the DAM approach has to be translated into a beneficial bidirectional data exchange that involve partly or fully the citizens of the area of interest. The user device can transmit the folk/technical data to the network, exchange positioning data, receive/send alarms. It has to be user-friendly, it could be embedded in other devices and it needs to support nav-com capabilities [4, 5].

In Table 2.1 the correspondence between DAM tools and supporting technologies is summarized.

Table 2.1. Possible supporting technologies for the seismic DAM tools.

Tool	Technology
Present and past technical parameters	WSN, BAN, PAN, PN, NF, IB
Present and past folk information	WSN, BAN, PAN, PN, NF, Semantic analyzers, IB
Data fusion and processing	Adaptive Artificial Intelligence, IB
Communications	D&S net
User devices	Integrated NavCom

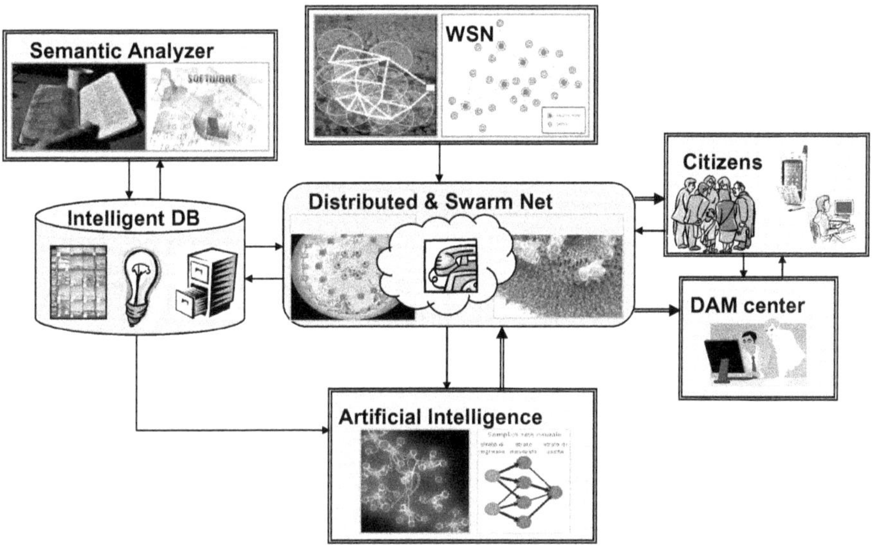

Fig. 2.2 The DAM ICT-based system concept.

Figure 2.2 reports the DAM system concept. A modeling of the phenomena and a seismic risk analysis are important steps behind the conceivement of the DAM process.

2.5 The Role of CTIF_Italy in the Seismic DAM

The Italian node of the Center for TeleInfrastructure (CTIF_Italy), created in 2006 at the University of Roma Tor Vergata, belongs to a worldwide network of research centers located in Denmark (headquarter), Japan, India and soon USA.

The CTIF network is based on a strong and effective interconnection among its nodes. As a consequence, the access for cooperation at institutional, industrial and scientific level on each node easily guarantees the access to the other ones. Each node promotes research activities, projects and education at local, national and international level.

The main mission of the CTIF network and, in particular, of CTIF_Italy deals with advanced architectures and applications for Information and Communication Technology (ICT): the key research areas include wireless communications, fixed infrastructures, satellite communications, satellite navigation, remote sensing, integrated networks and applications, extremely high frequency links.

Moving from the maturity of some ICT areas, CTIF_Italy is developing its research activities following two major goals.

The first goal derives from the idea that a research center can play a key-role in the process of translating an advanced concept into a successful product. The above process is complex, sometimes all but straightforward due to intense bureaucracy, shortage of public or private funding, sudden changes in the political or management scenarios.

In order to enhance the probability that the process be able to come to an effective completion, a research center should try to push its contribution in the process itself as far as its scientific and technical capabilities allow. The above aim can be met if the research center is keen to hardware and software development as well as to system conceivement and project management.

The CTIF network is the ideal environment for a research center to enhance those application and industry-prone capabilities that drive effectively the mentioned concept-to-product translation process.

In the above framework, CTIF_Italy has created a prototype Laboratory named ***HASCON*** (**Hardware an Algorithmic Solutions of Communication and Navigation**).

The other goal moves from ICT maturity. ICT is therefore ready to cross-fertilize other areas where its technological, modeling and architectural results can be effectively utilized.

Many research fields directly related with the improvement of the QoL of human beings can take great advantage from the maturity of ICT.

CTIF_Italy is addressing deep efforts in the use of ICT for improving the QoL, through an interdisciplinary research that, so far, deals with fields like Energy systems, Biotechnology (in particular stem cells systems) and DAM.

Fig. 2.3 Swarm network controller.

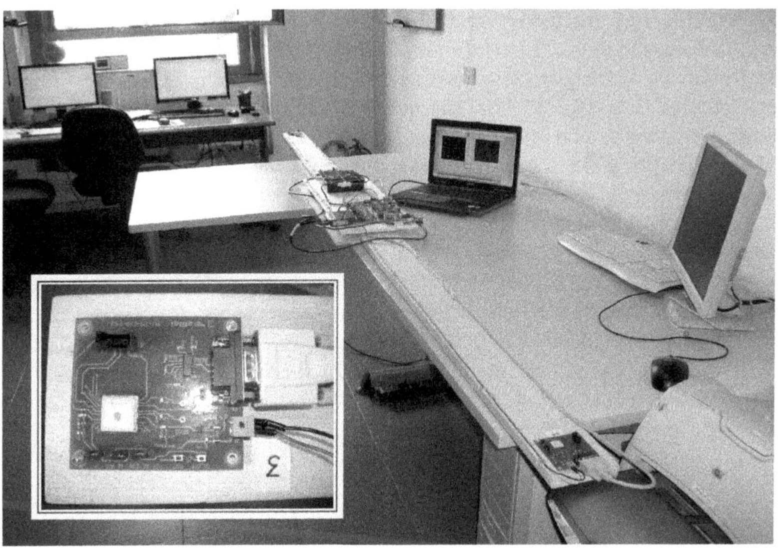

Fig. 2.4 GPS-based compass.

Two navcom prototypes developed in the CTIF_Italy *HASCON* laboratory are reported in Figures 2.3 and 2.4 that refer, respectively, to a control device of a swarm networks [6, 7] and to a GPS-based compass.

Due to the heritage in wireless systems, hardware and software navcom devices and applications, intelligent database for biotechnology systems,

CTIF_Italy role in the seismic DAM process is particularly focused on the following topics:

- WSN, BAN, PAN, PN
- swarm networks
- user navcom devices
- intelligent database,

contributing thus to the design and development of the DAM communication D&S network, the data collection system and the user terminals.

References

[1] R. Prasad and M. Ruggieri, "Technology Trends in Wireless Communications," *Artech House*, Boston, 2003, ISBN 1-58053-352-3.

[2] R. Prasad and M. Ruggieri, "Applied Satellite Navigation using GPS, GALILEO, and Augmentation Systems," *Artech House*, Boston, 2005 (ISBN: 1-58053-814-2).

[3] M. De Sanctis, S. Quaglieri, E. Cianca, and M. Ruggieri, "Energy Efficiency of Error Control for High Data Rate WPAN," *Wireless Personal Communications* (*Springer*), Special Issue on "*Advances on Wireless LANs and PANs*", July 2005, vol. 34, nos. 1–2, pp. 189–209.

[4] M. Ruggieri, "Next Generation of Wired and Wireless Networks: the NavCom Integration", *Springer Wireless Personal Communications*, June 2006, vol. 38, no. 1, pp. 79–88.

[5] E. Del Re and M. Ruggieri (Eds), "Satellite Communications and Navigation Systems," *Springer*, 2008, ISBN: 978-0-387-47522-6; e-ISBN: 978-0-387-47524-0.

[6] S. Barbera, S. Cacucci, F. Fedi, M. Ruggieri, G. Savarese, and C. Stallo, "A Geo-referenced Swarm Agent Enabling System: theory and demo application," *Proceedings ISMS*, Liverpool, January 2010, ISBN 978-0-7695-3973-7.

[7] S. Barbera, S. Cacucci, F. Fedi, M. Ruggieri, G. Savarese, and C. Stallo, "A Georeferenced Swarm Agent Enabling System for Hazardeous Application," *Proceedings UKSIM*, Cambridge, March 2010.

3

Artificial Adaptive Systems: Philosophy, Mathematics and Applications

Massimo Buscema

Semeion Research Center, Rome, Italy

3.1 Artificial Adaptive Systems

Artificial Adaptive Systems (AAS) form part of the vast world of **Natural Computation** (NC). **Natural Computation** (NC) is a subset of the **Artificial Sciences** (AS).

Artificial Sciences means those sciences for which an understanding of natural and/or cultural processes is achieved the recreation of those processes through automatic models.

In the AS, the computer is what writing represents for **natural language**: the AS consist of **formal algebra** for the generation of **artificial models** (structures and processes), in the same way in which natural languages are made up of semantics, syntax and pragmatics for the generation of **texts**.

In natural languages, writing is achieving **independence of the word from time**, through **space**; in the AS, the computer is the achievement of **independence of the model from the subject**, through automation.

Exactly as through writing a natural language can create **cultural objects** that were unthinkable before writing (stories, legal texts, manuals, etc.), in the same way the AS can create through the computer **automatic models** of unthinkable complexity.

Natural languages and Artificial Sciences, without writing and the computer, are therefore limited. But a writing not based on a natural language, or an automatic model not generated by formal algebra, are a set of scribbles (Figure 3.1).

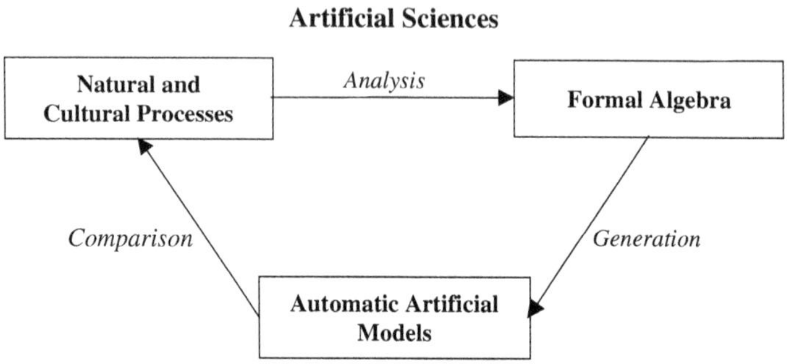

Fig. 3.1 The diagram shows how the analysis of Natural and/or Cultural Processes, that need to be understood, starts from a theory which, adequately formalized (Formal Algebra), is able to generate Automatic Artificial Models of those Natural and/or Cultural Processes. Lastly, the generated Automatic Artificial Models must be compared with the Natural and/or Cultural Processes of which they profess to be the model and the explanation.

In the AS, the understanding of any natural and/or cultural process occurs in a way that is proportional to the capacity of the automatic artificial model to recreate that process. The more positive the outcome of a comparison between original process and model generated is, the more likely it is the artificial model has explained the functioning rules of the original process.

However, this comparison cannot be made simple-mindedly. Sophisticated analysis tools are needed to make a reliable comparison between original process and artificial model.

Most of the analysis tools useful for this comparison consist of comparing the dynamics of the original process and the dynamics of the artificial model when the respective conditions in the surroundings are varied.

In sum, it could be argued that:

1. on varying the conditions in the surroundings, **the more** varieties of response dynamics are obtained both in the original process and in the artificial model; **and**
2. **the more** this dynamics between original process and artificial model is homologous, **therefore**
3. **the more** probable it is that the artificial model is a good explanation of the original process.

In Figure 3.2, we propose a taxonomic tree for characterisation of the disciplines that, through Natural Computation and Classic Computation, make up the Artificial Sciences system.

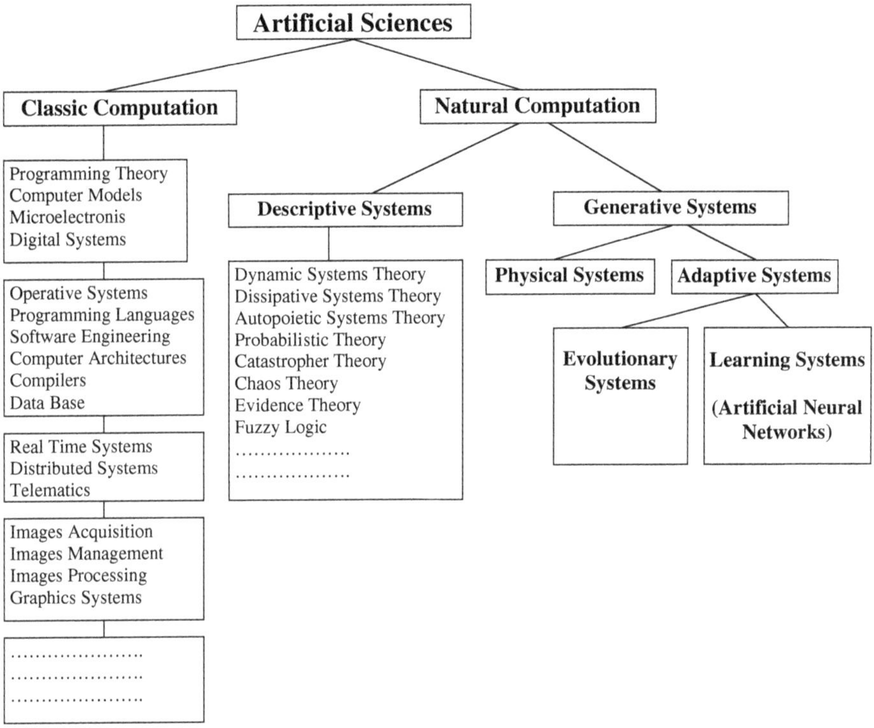

Fig. 3.2 Taxonomic tree of the disciplines that make up the Artificial Sciences system.

Natural Computation (NC) means that part of the Artificial Sciences (AS) that tries to construct automatic models of Natural and/or Cultural Processes through the local interaction of non-isomorphic microprocesses. In NC, it is therefore assumed that:

1. any process is the more or less contingent result of more basic processes tending to self-organise in time and space;
2. none of the microprocesses is in itself informative concerning the function that it will assume with respect to the others, nor the global process of which it will be part.

This computational philosophy, very economic for the creation of simple models, can be used effectively to create any type of process or model that is inspired by complex processes, or by processes regarding which the classic philosophies came up against considerable problems.

NC in fact deals with the construction of artificial models not simulating the complexity of Natural and/or Cultural Processes through rules, but through commitments that, depending on the space and time through which the process takes form, autonomously create a set of contingent and approximate rules.

NC does not try to recreate Natural and/or Cultural Processes by analysing the rules, through which it is wanted to make them function, and formalizing them into an artificial model. On the contrary, NC tries to recreate Natural and/or Cultural Processes by constructing artificial models able to create local rules dynamically, capable of change in accordance with the process itself.

The links that enable NC models to generate rules dynamically are similar to the Kantian transcendental rules: these are rules that establish the conditions of possibility of other rules.

In NC, a dynamic such as **learning to learn** is implicit in the artificial models themselves, whilst in Classical Computation it needs further rules.

The following are part of Natural Computation:

- **Descriptive Systems** (DS): these are those disciplines that have developed, whether or not intentionally, **formal algebra** that has

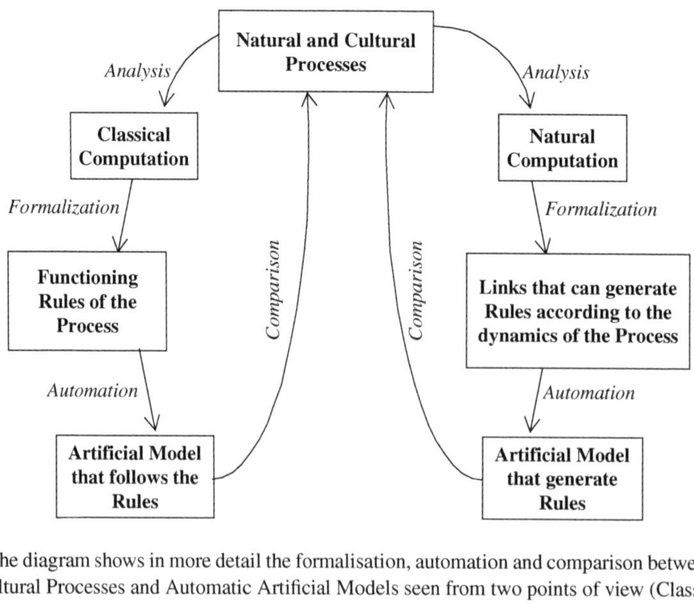

Fig. 3.3 The diagram shows in more detail the formalisation, automation and comparison between Natural and/or Cultural Processes and Automatic Artificial Models seen from two points of view (Classical Computation and Natural Computation). Each point of view can be seen as a cycle that can repeat itself several times. This allows to deduce that the human scientific process characterising both the cycles, resembles more the Natural Computation than the Classical Computation one.

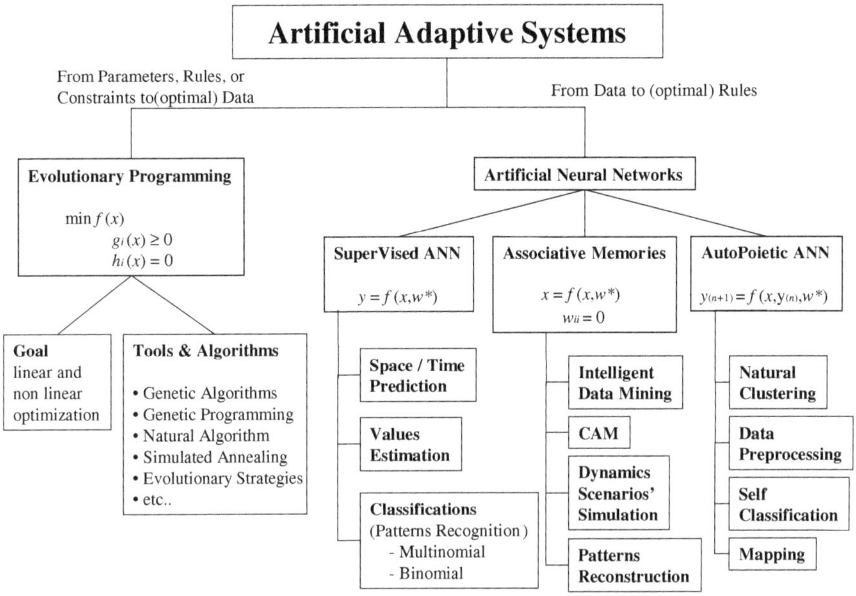

Fig. 3.4 Artificial Adaptive Systems — general diagram.

proved particularly effective in drawing up appropriate functioning links of artificial models generated within NC (for example: the Theory of the Dynamic Systems, the Theory of Autopoietic Systems, Fuzzy Logic, etc.).

- **Generative Systems** (GS): these are those theories of NC that have explicitly provided formal algebra aimed at generating artificial models of Natural and/or Cultural Processes through links that create dynamic rules in space and in time.
- In turn, Generative Systems can be broken down into:
- **Physical Systems** (PS): this means grouping those theories of Natural Computation whose **generative algebra** creates artificial models comparable to physical and/or cultural processes, only when the artificial model reaches given evolutive stages (limit cycles type). Whilst not necessarily the route through which the links generate the model is itself a model of the original process. In brief, in these systems the generation time of the model is not necessarily an artificial model of evolution of the process time (for example: Fractal Geometry, etc.).

- **Artificial Adaptive Systems** (AAS): this means those theories of Natural Computation whose generative algebra creates artificial models of Natural and/or Cultural Processes, whose birth process is itself an artificial model comparable to the birth of the original process. They are therefore theories assuming the emergence time of the model as a formal model of the process time itself.
- In short: for these theories, each phase of artificial generation is a model comparable to a natural and/or cultural process.
- Artificial Adaptive Systems in turn comprise:
- **Learning Systems (Artificial Neural Networks** — ANNs): these are algorithms for processing information that allow to reconstruct, in a particularly effective way, the approximate rules relating a set of "explanatory" data concerning the considered problem (the Input), with a set of data (the Output) for which it is requested to make a correct forecast or reproduction in conditions of incomplete information.
- **Evolutionary Systems** (ES): this means the generation of adaptive systems changing their architecture and their functions over time in order to adapt to the environment into which they are integrated, or comply with the links and rules that define their environment and, therefore, the problem to be simulated. Basically, these are systems that are developed to find data and/or optimum rules within the statically and dynamically determined links and/or rules.
- The development of a genotype from a time t_i to a time $t_{(i+n)}$ is a good example of the development over time of the architecture and functions of an adaptive system.

3.2 A Brief Introduction to Artificial Neural Networks

3.2.1 Architecture

ANNs are a family of methods stimulated by the workings of the human brain. Currently ANNs comprise a range of very different models, but they all share the following characteristics:

- The fundamental elements of each ANN are the **Nodes**, also known as Processing Elements (PE), and their **Connections**.

- Each node in an ANN has its own **Input**, through which it receives communications from the other nodes or from the environment; and its own **Output**, through which it communicates with other nodes or with the environment. Finally it has a function, $f(\cdot)$, by which it transforms its global input into output.
- Each Connection is characterized by the force with which the pair of nodes excite or inhibit each other: positive values indicate excitatory connections, negative ones indicate inhibitory connections.
- Connections between nodes may change over time. This dynamic triggers a **learning process** throughout the entire ANN. The way (the law by which) the connections change in time is called the "Learning Equation".
- The overall dynamic of an ANN is linked to **time**: in order to change the connections of the ANN properly, the environment must act on the ANN several times.
- When ANNs are used to process data, these latter are their environment. Thus, in order to process data, these latter must be subjected to the ANN several times.
- The overall dynamic of an ANN depends exclusively on the local interaction of its nodes. The final state of the ANN must, therefore, evolve 'spontaneously' from the interaction of all of its components (nodes).
- Communications between nodes in every ANN tend to occur in **parallel**. This parallelism may be **synchronous** or **asynchronous** and each ANN may emphasize it in a different way. However, any ANN **must** present some form of parallelism in the activity of its nodes.
- From a theoretical viewpoint this parallelism does not depend on the hardware on which the ANNs are implemented.

Every ANN must present the following architectural components:

- Type, number of nodes and their properties
- Type, number of connections and their location
- Type of Signal Flow Strategy
- Type of Learning Strategy.

3.2.2 The Nodes

There can be **three types** of ANN nodes, depending on the position they occupy within the ANN.

- Input nodes: these are the nodes that (also) receive signals from the environment outside the ANN.
- Output nodes: these are the nodes whose signal (also) acts on the environment outside the ANN.
- Hidden nodes: these are the nodes that receive signals only from other nodes in the ANN and send their signal only to other nodes in the ANN.

The number of input nodes depends on the way the ANN is intended to **read** the environment. The input nodes are the ANN's **Sensors**. When the ANN's environment consists of data the ANN should process, the input node corresponds to a sort of data **variable**.

The number of output nodes depends on the way one wants the ANN to **act** on the environment. The output nodes are the **Effectors** of the ANN. When the ANN's environment consists of data to process, the output nodes represent the variables sought or the results of processing.

The number of hidden nodes depends on the complexity of the function one intends to map between the input nodes and the output nodes.

The nodes of each ANN may be grouped into classes of nodes sharing the same properties.

Normally these classes are called layers.

Various types can be distinguished:

- MonoLayer ANNs: all nodes of the ANN have the same properties.
- MultiLayer ANNs: the ANN nodes are grouped in functional classes; e.g.,: nodes (a) sharing the same signal transfer functions; (b) receiving the signal only from nodes of other layers and send them only to new layers; etc.
- Nodes Sensitive ANNs: each node is specific to the position it occupies within the ANN; e.g., the nodes closest together communicate more intensely than they do with those further away.

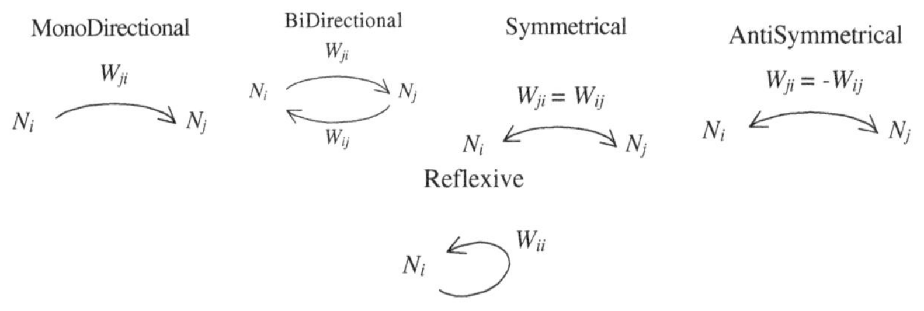

Fig. 3.5 Types of possible connections.

3.2.3 The Connections

There may be various types of connections: MonoDirectional, BiDirectional, Symmetrical, AntiSymmetrical and Reflexive:

The number of connections is proportional to the memory capabilities of the ANN. Positioning the connections may be useful as methodological preprocessing for the problem to be solved, but it is not necessary. An ANN where the connections between nodes or between layers are not all enabled is called an ANN with **Dedicated Connections**; otherwise it is known as a **maximum gradient** ANN.

In each ANN the connections may be:

- Adaptive: they change depending on the learning equation.
- Fixed: the remain at fixed values throughout the learning process.
- Variable: they change deterministically as other connections change.

3.2.4 The Signal Flow

In every ANN the signal may proceed in a direct fashion (from input to output) or in a complex fashion.

Thus we have two types of Flow Strategy:

- **Feed Forward ANN**: the signal proceeds from the input to the output of the ANN passing all nodes only once.
- **ANN with Feedback**: the signal proceeds with specific feedbacks, determined beforehand, or depending on the presence of particular conditions.

The ANNs with Feedback are also known as **Recurrent ANNs**, and are the most plausible from a biological point of view. They are often used to process timing signals, and they are the most complex to deal with mathematically.

In an industrial context, therefore, they are often used with feedback conditions determined a priori (in order to ensure stability).

3.3 The Learning

Every ANN can learn over time the properties of the environment in which it is immersed, or the characteristics of the data which it presents, basically in one of 2 ways (or mixture of both):

- By reconstructing approximately the probability density function of the data received from the environment, compared with preset constraints.
- By reconstructing approximately the parameters which solve the equation relating the input data to the output data, compared with preset constraints.

The first method is known in the context of ANNs as **Vector Quantization**; the second method is **Gradient Descent**. The Vector Quantization method articulates the input and output variables in **hyperspheres** of a defined range. The Gradient Descent method articulates the input and output variables in **hyperplanes**.

The difference between these two methods becomes evident in the case of a Feed Forward ANN with at least one hidden unit layer. With Vector Quantization the hidden units encode **locally** the more relevant traits of the input vector.

At the end of the learning process, each hidden unit will be a **Prototype** representing one or more relevant traits of the input vector in definitive and exclusive form.

With Gradient Descent, the hidden units encode **in a distributed manner** the most relevant characteristics of the input vector.

At the end of the learning process, each hidden unit will tend to represent the relevant traits of the input in a fuzzy and non-exclusive fashion.

Summing up: the Vector Quantization develops a **local** learning, the Gradient Descent develops a **distributed** or **vectorial** learning.

Considerable differences exist between the two approaches:

- Distributed learning is computationally more efficient than local learning. It may also be more plausible biologically (not always or in every case).
- When the function that connects input to output is non linear, distributed learning may "jam" on local minimums due to the use of the Gradient Descent technique
- Local learning is often quicker than distributed learning.
- The regionalization of input on output is more sharply defined using Vector Quantization than when using Gradient Descent.
- When interrogating an ANN trained with Vector Quantization, the ANN responses cannot be different from those given during learning; in the case of an ANN trained with Gradient Descent the responses may be different from those obtained during the learning phase.
- This feature is so important that families of ANNs treating the signal in 2 steps have been designed: first with the Quantization method and then with the Gradient method.
- Local learning helps the researcher to understand how the ANN has interpreted and solved the problem; distributed learning makes this task more complicated (though not impossible).
- Local learning is a competitive type, distributed learning presents aspects of both competitive and cooperative behavior between the nodes.

3.4 Artificial Neural Networks Typology

ANNs may in general be used to resolve basically 3 types of complex problems and consequently they can be classify into three sub families.

3.4.1 Supervised ANNs

The first type of problem that an ANN can deal with can be expressed as follows: *given N variables, about which it is easy to gather data, and M variables, which differ from the first and about which it is difficult and costly to gather data, assess whether it is* ***possible to predict*** *the values of the M variables on the basis of the N variables.*

These family of ANNs are named **Supervised** ANNs (**SV**) and their prototypical equation is:

$$y = f(x, w^*) \tag{3.1}$$

where y is the vector of the M variables to predict and/or to recognize(target), x is the vector of N variables working as networks inputs, w is the set of parameters to approximate and $f(\cdot)$ is a non linear and composed function to model.

When the M variables occur subsequently in time to the N variables, the problem is described as a **prediction** problem; when the M variables depend on some sort of typology, the problem is described as one of **recognition** and/or **classification**.

Conceptually it is the same kind of problem: *using values for some **known variables** to predict the values of other **unknown variables**.*

To correctly apply an ANN to this type of problem we need to run a **validation protocol**.

We must start with a good sample of cases, in each of which the N variables (known) and the M variables (to be discovered) are both **known** and **reliable**.

The sample of complete data is needed in order to:

- *train* the ANN, and
- *assess* its predictive performance

The validation protocol uses **part** of the sample to **train** the ANN (Training Set), whilst the **remaining cases** are used to assess the **predictive capability** of the ANN (Testing Set or Validation Set).

In this way we are able to test the reliability of the ANN in tackling the problem *before* putting it into operation.

3.4.2 Dynamic Associative Memories

The second type of problem that an ANN can be expressed as follows: *given N variables defining a dataset, find out its optimal connections matrix able to define each variable in terms of the others and consequently to approximate the hyper-surface on which each data-point is located.*

This second sub-family of ANNs is named **Dynamic Associative Memories (DAM)**.

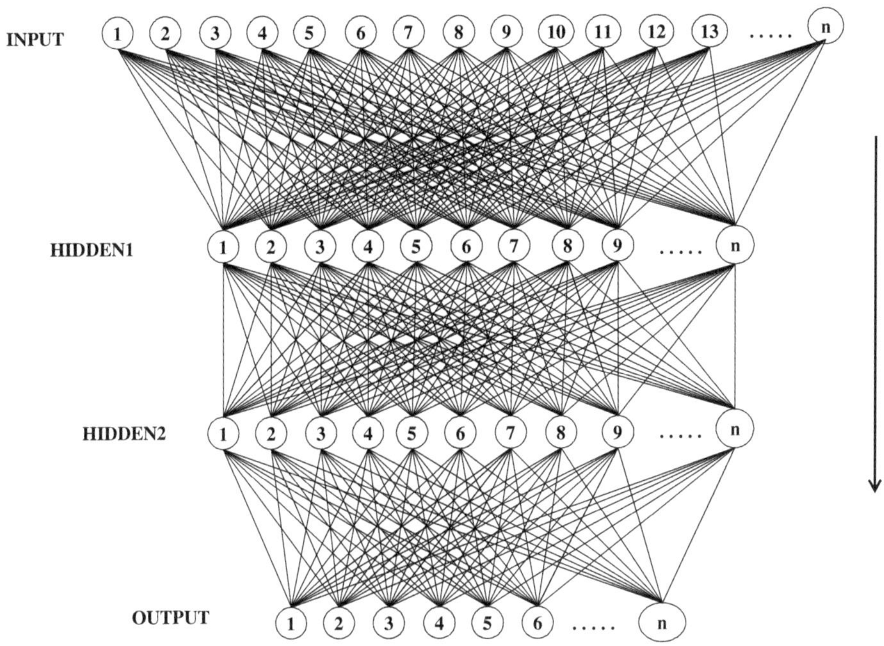

Fig. 3.6 Example of Supervised ANN.

The specificity of these ANNs is incomplete patterns reconstruction, dynamic scenarios simulation and possible situations prototyping.

Their representative equation is:

$$x^{n+1} = f(x^n, w^*) \tag{3.2}$$

Where x^n is the N variables evolution in internal time of ANN, w^* is connections matrixes approximating the parameters of the hyper-surface representing the dataset, and $f(\cdot)$ is some suitable non linear and eventually composed function governing the process.

DAM ANNs after the training phase need to be submitted to a validation protocol named "Data Reconstruction Blind Test". In this test the capability of a DAM Ann to rebuild complete data from uncompleted ones is evaluated form a quantitative point of view.

3.4.3 Autopoietic ANNs

The third type of ANNs can be describe as follow: *given N variables defining M records in a dataset, evaluate how these variables are distributed and how*

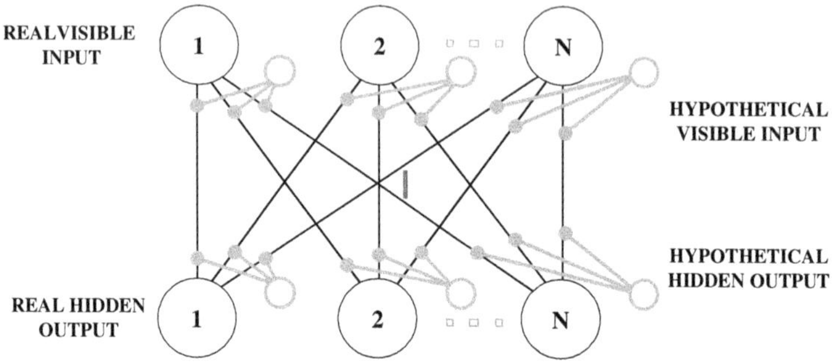

Fig. 3.7 Example of Dynamic Associative Memory — New Recirculation ANN.

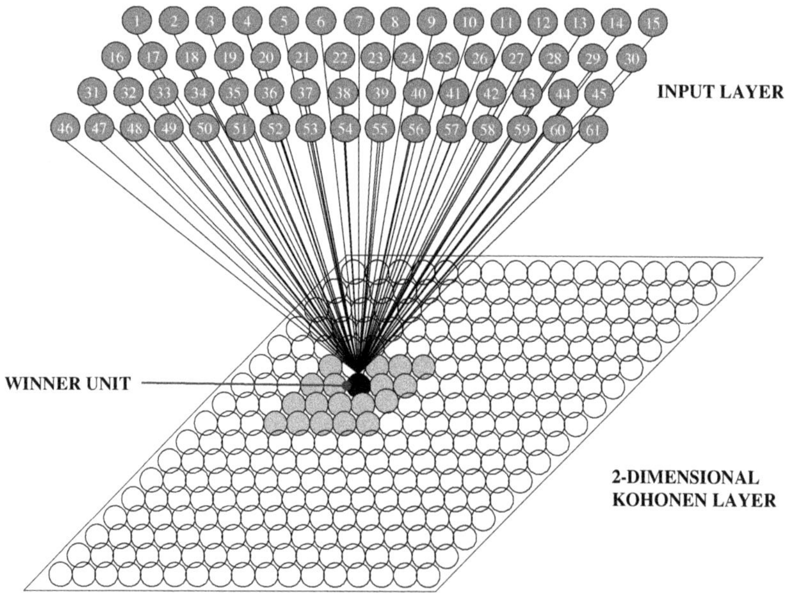

Fig. 3.8 Example of Unsupervised ANN for natural clustering — Self-Organizing Map.

these records are naturally clustered in a small projection space $K(K \ll N)$ according to their most important relationships.

These ANNs are named Unsupervised or also Auto-Poietic ANNs (US). Their specificity is the non linear extraction of the similarities among records in a database, using all the variables at the same time. One important feature of these ANNs is also the possibility that some of them have to visualize in a 2

Table 3.1. Semantics and Syntax of ANNs.

Summary Table

1. Table 1	Type	Dynamic	Properties
nodes	• Input • Output • Hidden	• Type of equation I → 0	• No Layer (each node is distinct from every other) • MultiLayer (each node is the same as those of its own layer) • MonoLayer (each node is the same as the others)
connections	• Symmetrical • AntiSymmetrical • MonoDirectional • BiDirectional • Reflexive	• Adaptive • Fixed • Variables	• Maximum connections • Dedicated connections
2. Table 2	Flow Strategy	Learning Strategy	
Type of ANN	• Feed Forward • with Parametric or Adaptive Feed Back • IntraNode • IntraLayer • Among Layers • Among ANNs	• Approximation of the function: • Gradient Descent • Vector Quantization • Learning conditions of the function: • Supervised • Dynamic Associative Memories • Unsupervised or AutoPoietic	

or 3 dimensional map the geographical similarities among records and among variables.

The prototypical equation of the US ANNs is:

$$y^{n+1} = f(y^n, x, w^*) \tag{3.3}$$

where y is the projection result along the time, x is the input vector (independent variables) and w is the set of parameters (codebooks) to be approximated.

In US ANNs, the codebooks (w) after the training phase represent an interesting case of cognitive abstraction: in each codebook the ANN tends to develop its abstract cognitive representation of some of the data which it learnt.

3.5 Artificial Neural Networks Application

The most typical problem that an ANN can deal with can be expressed as follows: given *N* variables, about which it is easy to gather data, and *M* variables, which differ from the first and about which it is difficult and costly to gather data, assess whether it is possible to predict the values of the *M* variables on the basis of the *N* variables.

When the *M* variables occur subsequently in time to the *N* variables, the problem is described as a prediction problem; when the *M* variables depend on some sort of static and/or dynamic factor, the problem is described as one of recognition and/or discrimination and/or extraction of fundamental traits.

To correctly apply an ANN to this type of problem we need to run a validation protocol.

We must start with a good sample of cases, in each of which the *N* variables (known) and the *M* variables (to be discovered) are both known and reliable.

The sample of complete data is needed in order to:

- train the ANN, and
- assess its predictive performance.

The validation protocol uses part of the sample to train the ANN (Training Set), whilst the remaining cases are used to assess the predictive capability of the ANN (Testing Set or Validation Set).

In this way we are able to test the reliability of the ANN in tackling the problem before putting it into operation.

Different types of protocol exist in literature, each presenting advantages and disadvantages. One of the most popular is employed the so called 5×2 cross-validation protocol [8] which produces 10 elaborations for every sample. It consists in dividing the sample five times in 2 different sub samples, containing similar distribution of cases and controls (Figure 3.9).

3.5.1 Description of the Standard Validation Protocol

The protocol from the point of view of a general procedure, consists of the following steps:

1. subdividing the DB in a random way into two sub-samples: the first named Training Set and the second called Testing Set;

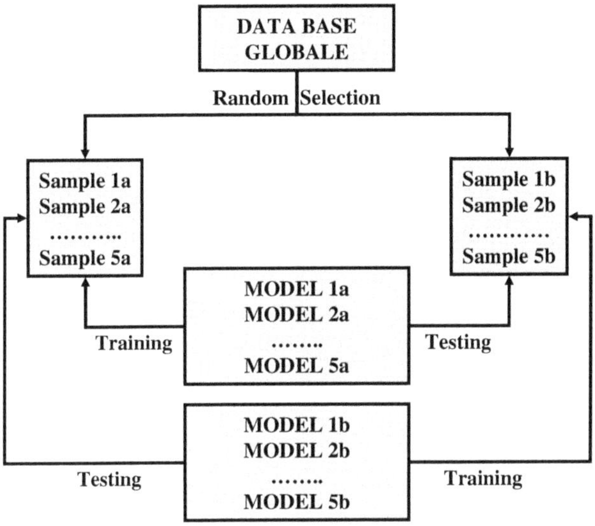

Fig. 3.9 The 5 × 2 cross-validation protocol.

2. choosing a fixed ANN, and/or another model, which is trained on the Training Set, in this phase the ANN learns to associate the input variables with those that are indicated as targets;
3. at the end of the training phase the weight matrix produced by the ANN is saved and frozen together with all of the other parameters used for the training;
4. with the weight matrix saved the Testing Set, which it has not seen before, is shown to the ANN so that in each case the ANN can express an evaluation based on the previously carried out training, this operation takes place for each input vector and every result (output vector) is not communicated to the ANN;
5. the ANN is in this way evaluated only in reference to the generalization ability that it has acquired during the Training phase;
6. a new ANN is constructed with identical architecture to the previous one and the procedure is repeated from point 1.

3.5.2 Artificial Neural Networks and Problems in Need of Solution

In theory the number of types of problems that can be dealt with using ANNs is limitless, since the methodology is broadly applicable, and the problems spring up as fast as the questions that society, for whatever reason, poses. So, let us remind ourselves of the criteria that must be satisfied for the adoption of an ANN to be worthwhile:

- The problem must be a complex one.
- Other methods have not produced satisfactory results.
- An approximate solution to the problem is valuable, not only a certain or best solution.
- An acceptable solution to the problem offers great savings in human and/or economic terms.
- There exists a large case history demonstrating the 'strange' behavior to which the problem pertains.

Figure 3.10 summarizes the conditions which best claim for neural networks analysis.

3.5.3 A Special Feature of Neural Networks Analysis: Variables Selection

ANNs are able to simultaneously handle a very high number of variables notwithstanding their underlying non linearity. This represent a tremendous advantage in comparison with classical statistics models in a situation in which the quantity of available information is enormously increased and non linearity dominates. With ANNs one is neither concerned about the actual number of variables nor about their nature. Due to their particular mathematic

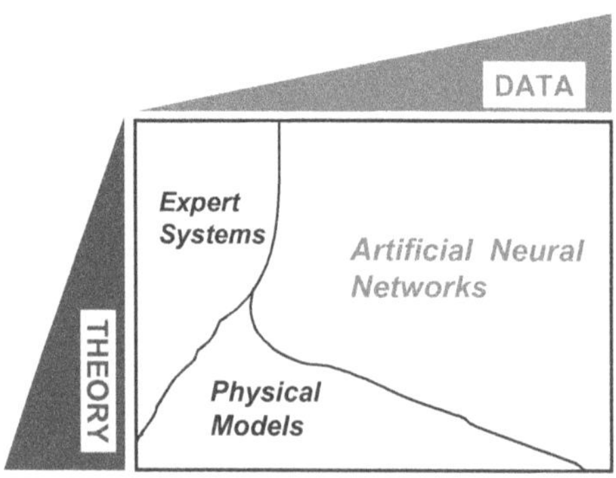

Fig. 3.10 Schematic comparison of Artificial Neural Network with other analysis techniques.

infrastructure, AAS have no limits in handling increasing amounts of variables which constitute the input vector for the recursive algorithms.

Discriminant analysis, logistic regression or other linear or semi-linear techniques typically employ a limited number of variables in building up their model: those with a prominent linear correlation with the dependent variable. In other words for them a specific criterium exists, i.e., correlation index, which indicates which of the variables available has to be used to build a predictive model addressed to the particular problem under evaluation.

ANNs, being a sort of universal approximation system, are able to use a wider range of information available, also with variables with a very poor linear correlation index. For this reason the natural approach is to include all of the variables that according to clinician experience may have a priori a connection with the problem being studied. When the ANNs are exposed to all these variables of the data set they can very easily actually approximate all the information available during the training phase. This is the strength but unfortunately also the weakness of ANNs.

In fact, almost inevitably a number of variables which do not contain specific information pertinent to the problem being examined are processed. These variables, inserted in the model, cause act as a sort of "white noise" and interfere with the generalization ability of the network in the testing phase. In this case the ANN lose some potential external validity in the testing phase with consequent reduction in the overall accuracy. This explain why, in absence of systems able to select appropriately the variables containing the pertinent information, the performance of ANNs might remain below expectations, being only slightly better than classical statistical procedures.

One of the most sophisticated and useful approach to overcome this limitation is to define the subgroup of variables to be selected with another family of Artificial Adaptive Systems: evolutionary algorithms (EAs).

3.5.4 Evolutionary Algorithms

At variance with neural networks which are adaptive systems able to discover the optimal hidden rules explaining a certain data set, the Evolutionary Algorithms (EA) are Artificial Adaptive Systems able to find optimal data when fixed rules or constraints have to be respected. They are, in other words, optimization tools which become fundamental when the space of possible states in a dynamic system tends toward infinitum. This is just the case of variables selection. Given a certain large amount of dichotomous variables (for example 100) the problem to define the most appropriate subgroup of them to better solve the problem under examination, has a very large space of possible states and exactly: 2^{100}. The computational time required to sort all possible variables subsets in order to submit them to ANNs processing would be in the order of million years; a so called NP (non polynomial) hard mathematical problem.

The introduction of variable selection systems generally results in a dramatic improvement of ANNs performance.

The Input Selection (IS.) is an example of the adaptation of an evolutionary algorithm to this problem .

This is a selection mechanism of the variables of a fixed dataset, based on the evolutionary algorithm GenD [5]. The IS system becomes operative on a population of ANNs, each of them with a capability of extract a different pool of independent variables. Through the GenD evolutionary algorithm, the different "hypotheses" of variable selection, generated by each ANNs, change over time, generation after generation. When the evolutionary algorithm no longer improves, the process stops, and the best hypothesis of the input variables is selected and employed on the testing subset. The goodness-of-fit rule of GenD promotes, at each generation, the best testing performance of the ANN model with the minimal number of inputs.

An example of an application of IS system is describes in the paper of Buscema et al. [6] containing also a theoretical description of the neuro-evolutionary systems.

References

[1] J. A. Anderson and E. Rosenfeld (eds.) (1988), *Neurocomputing Foundations of Research*, The MIT Press, Cambridge, MA, 1988.

[2] M. A. Arbib (ed.), *The Handbook of Brain Theory and Neural Networks, A Bradford Book*, The MIT Press, Cambridge, Massachusetts, London, England, 1995.

[3] M. Buscema et al., Artificial Neural Networks and Complex Social Systems, vol. 33(1): Theory, vol. 33(2): Models, vol. 33 (3): Applications, Substance Use & Misuse, Marcel Dekker, New York, 1998.

[4] M. Buscema and Semeion Group, Reti Neurali Artificiali e Sistemi Sociali Complessi, vol. I: Teoria e Modelli, vol. II: Applicazioni, Franco Angeli, Milano, 1999.

[5] M. Buscema, Genetic doping algorithm (GenD): theory and applications. Expert Systems, 2004; 21: 63–79.

[6] M. Buscema, E. Grossi, M. Intraligi, N. Garbagna, A. Andriulli, M. Breda, An optimized experimental protocol based on neuro-evolutionary algorithms: application to the classification of dyspeptic patients and to the prediction of the effectiveness of their treatment. *Artificial Intelligence Med*, 2005; 34: 279–305.

[7] G. A. Carpenter and S. Grossberg, Pattern Recognition by Self-Organizing Neural Network, MIT Press, Cambridge, MA, 1991.

[8] T. G. Dietterich, Approximate statistical tests for comparing supervised classification learning algorithms. *Neural Computing*, 1998; 7:1895–1924.

[9] S. Grossberg, How Does the Brain Build a Cognitive Code? *Psychological Review*, 87, 1978.

[10] J. J. Hopfield, Neural networks and physical systems with emergent collective computational abilities, Proceedings of the National Academy of Sciences 79, in Anderson 1988.

[11] J. J. Hopfield, Neurons with graded response have collective computational properties like those of two-state neurons, Proceedings of the National Academy of Sciences USA, Bioscience 81, 1984.

[12] J. J. Hopfield and D. W. Tank, Neural Computation of Decisions in Optimization Problems, *Biological Cybernetics*, 52, 1985.

[13] J. J. Hopfield and D. W. Tank, Computing With Neural Circuits: A Model, Articles Science, vol. 233, 8 August 1986.

[14] T. Kohonen, Correlation matrix memories, IEEE Transactions on Computers C-21, in Anderson 1988.

[15] M. Minsky, *Neural Nets and the Brain-Model Problem, Doctoral Dissertation*, Princeton University, 1954.

[16] M. Minsky and S. Papert. *Perceptrons*, The MIT Press, Cambridge, MA, (extended edition, 1988).

[17] B. Reetz, Greedy solution to the Travelling Sales Person Problem, ATD, vol. 2, May, 1993.

[18] F. Rosenblatt, *Principles of Neurodynamics*, Spartan, New York, 1962.

[19] D. E. Rumelhart and J. L. Mcclelland (eds.), Parallel Distributed Processing, vol. 1 Foundations, Explorations in the Microstructure of Cognition, vol. 2 Psychological and Biological Models. The MIT Press, Cambridge, MA, London, England 1986.

[20] P. Werbos, Beyond Regression: New Tools for Prediction and Analysis in Behavioral Sciences, Phd Thesis, Harvard, Cambridge, MA, 1974.

Part II

Modelling and Tools

4

Quakes Prediction Using Highly Non Linear Systems and A Minimal Dataset

Massimo Buscema* and Roberto Benzi†

*Semeion Research Center, Rome, Italy
†University of Tor Vergata, Rome, Italy

4.1 Introduction: Earthquakes, Data and Mathematical Algorithms

The earthquakes predictability is an important issue for the scientific community and the whole human race. Geophysicists and geologists have been debating the predictability of earthquakes for a long time (Kagan, 1997; Kagan and Jackson, 2000), but they seem far from having reached an agreement.

Earthquakes dynamics seems to be highly non linear and non stationary. In other words, the process of generation of earthquakes has proven to be too complex to be tackled through traditional linear statistics.

Artificial Neural Networks (ANNs) are the classic algorithms to deal with the management of complex data. But in a recent review about earthquakes prediction, no one paper using ANNs is quoted (Kanamori, 2003).

Nevertheless, in the last years many scientists tried to apply ANNs to the issues concerning earthquakes, obtaining interesting and promising results (Sharma and Arora, 2005; Ashif et al., 2007; Suratgar et al., 2008).

In this paper we intend to analyze the possibility to estimate earthquakes magnitude starting from quite elementary data. Our aims are the following:

a. To demonstrate, from the experimental viewpoint, that also a simple dataset on earthquakes presents a rich information structure, one that is 'invisible' to the traditional linear algorithms;

Table 4.1.

	Magnitude	
	R	R2
Year	−0.2638	0.0696
Month	0.0036	0.0000
Day	−0.0032	0.0000
Time	0.0010	0.0000
Latitude	−0.4266	0.1820
Longitude	0.2408	0.0580
Depth	0.2597	0.0674

b. To suggest that earthquakes dynamics is a global phenomenon, that is hard to understand relying only upon regional data;
c. To flesh out some cyclic, time independent patterns of earthquakes dynamics.

4.2 A Simple Dataset

We have collected data about earthquakes over the world from 1976 to 2002 (Houston University, Data Mining and Machine Learning Group: http://www.tlc2.uh.edu/dmmlg/Datasets/earth_data.htm. and USGS: http://earthquake.usgs.gov/.

These data regard 324541 events whose magnitude is bigger than 0.1. Each event has 8 attributes.

Year, Month, Day, Time, Latitude, Longitude, Depth and Magnitude.

The linear correlation between each attribute and the Magnitude is statistically non relevant (see Table 4.1), the statistical independence among the variables is high (see Table 4.3), and the frequency distribution of the Magnitudes into the whole sample seems to be at least bi-modal (see Figure 4.1 and Table 4.2).

4.3 The Validation Protocol and The Error Measure

To measure the capability of different algorithms to estimate the Magnitude of each event using as independent variables the 7 attributes already presented (Year, Month, Day, Time, Latitude, Longitude and Depth), we have formulated a classic Training and Testing protocol: the whole sample was divided

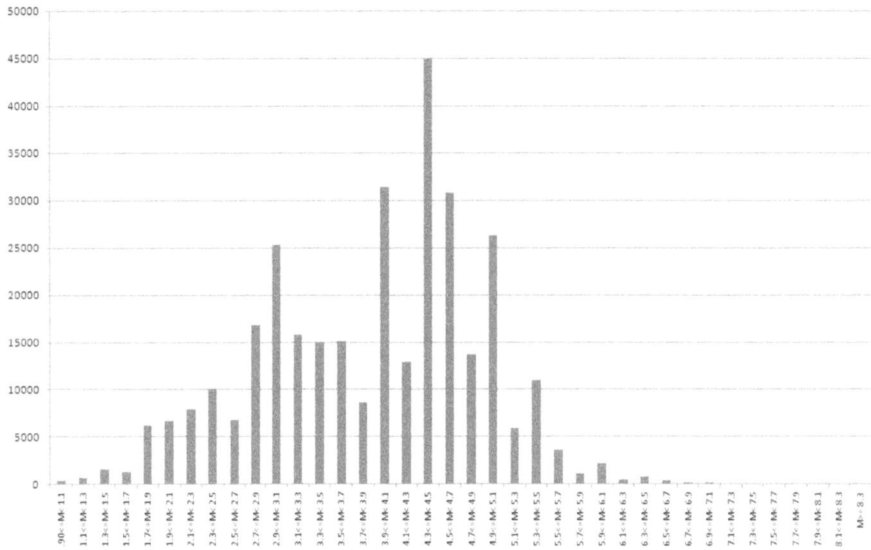

Fig. 4.1 Magnitude distribution.

randomly into two halves (training set and testing set) for each experimentation. Considering the huge amount of data available, this simple procedure is statistically effective.

All the experimentations we present will not consider time sequentially among events: each event is assumed as a possible state of a combinatorial system. We intend to establish whether a cyclic, time independent information structure, is buried into the dataset. Consequently, we will not use in our experimentations Recurrent ANNs.

Further, our experimentations are divided into two types:

1. Function Approximation: the algorithm, trained on the training set, will try to estimate the real Magnitude of each event of the testing set, using only the seven, already described, independent variables;
2. Classification: the algorithm, trained in the same way, will try to classify each event of the testing set into a predefined number of classes, each one representing a specific Magnitude interval.

To measure the mean error of each algorithm in the testing set, we will use different types of cost functions (in the following equations we assume we

Table 4.2.

Bins (M = Magnitude)	Number of Events
10 <= M < 0.30	2
30 <= M < 0.50	5
50 <= M < 0.70	27
70 <= M < 0.90	54
90 <= M < 1.1	434
1.1 <= M < 1.3	725
1.3 <= M < 1.5	1589
1.5 <= M < 1.7	1304
1.7 <= M < 1.9	6171
1.9 <= M < 2.1	6673
2.1 <= M < 2.3	7897
2.3 <= M < 2.5	10062
2.5 <= M < 2.7	6827
2.7 <= M < 2.9	16837
2.9 <= M < 3.1	25314
3.1 <= M < 3.3	15840
3.3 <= M < 3.5	15013
3.5 <= M < 3.7	15098
3.7 <= M < 3.9	8582
3.9 <= M < 4.1	31376
4.1 <= M < 4.3	12949
4.3 <= M < 4.5	44952
4.5 <= M < 4.7	30846
4.7 <= M < 4.9	13704
4.9 <= M < 5.1	26317
5.1 <= M < 5.3	5860
5.3 <= M < 5.5	10972
5.5 <= M < 5.7	3561
5.7 <= M < 5.9	1119
5.9 <= M < 6.1	2166
6.1 <= M < 5.3	458
6.3 <= M < 6.5	813
6.5 <= M < 6.7	349
6.7 <= M < 6.9	234
6.9 <= M < 7.1	165
7.1 <= M < 7.3	52
7.3 <= M < 7.5	95
7.5 <= M < 7.7	41
7.7 <= M < 7.9	32
7.9 <= M < 8.1	9
8.1 <= M < 8.3	10
M >= 8.3	7

Table 4.3. Mutual Information.

	Year	Month	Day	Time	Latitude	Longitude	Depth	Magnitude
Year	0.449446	0.426062	0.398164	0.434928	0.488728	0.459655	0.621256	0.512036
Month	0.426062	0.402678	0.374779	0.411544	0.465343	0.436271	0.597872	0.488652
Day	0.398164	0.374779	0.346881	0.383646	0.437445	0.408372	0.569973	0.460753
Time	0.434928	0.411544	0.383646	0.42041	0.474209	0.445137	0.606738	0.497518
Latitude	0.488728	0.465343	0.437445	0.474209	0.528009	0.498936	0.660537	0.551317
Longitude	0.459655	0.436271	0.408372	0.445137	0.498936	0.469863	0.631464	0.522244
Deep	0.621256	0.597872	0.569973	0.606738	0.660537	0.631464	0.793065	0.683845
Magnitude	0.512036	0.488652	0.460753	0.497518	0.551317	0.522244	0.683845	0.574626

have only one dependent variable):

1. The Root Mean Square Error (RMSE), a traditional measure for ANNs:

$$
\begin{aligned}
Rmse &= \sqrt[2]{\frac{\frac{1}{2} \cdot \sum_k^M (t_k - y_k)^2}{M}};\\
M &= \text{Records Number};\\
t_k &= \text{the k-th real magnitude}; \quad t \in 0, 1;\\
y_k &= \text{the k-th predicted magnitude}; \quad y \in 0, 1.
\end{aligned}
$$

2. The Linear Correlation Index (LC):

$$
\begin{aligned}
LC &= \frac{\sum_{k=1}^M (t_k - \bar{t}) \cdot (y_k - \bar{y})}{\sqrt{\sum_{k=1}^M (t_k - \bar{t})^2 \cdot \sum_{k=1}^M (y_k - \bar{y})^2}};\\
-1 &\leq LC \leq +1.
\end{aligned}
$$

3. The Absolute Mean Error (AbsError):

$$
\begin{aligned}
AbsErr &= F\left(\frac{\sum_k^M |t_k - y_k|}{M}\right);\\
F() &= \text{Linear Function to re-scale the error into}\\
&\quad\ \text{the original interval of Magnitude}
\end{aligned}
$$

4. The Weighted Error (Tau):

$$Tau = -\sum_{k}^{M} \frac{(t_k - y_k)^2}{2 \cdot \sigma^2}$$

$$-\infty \leq Tau \leq 0;$$

$$\sigma^2 = Variance$$

Each one of the presented cost functions has its own limits and advantages, although in most cases they are strongly correlated. We believe that the Weighted Error (Tau) cost function is the most effective, because it relates the square error to the variance of the testing set.

4.4 Function Approximation: The Training Phase

We have divided randomly the dataset into two halves: the training set (167271 events) and the testing set (167270 events).

We have trained 5 different algorithms on the same data:

a. Linear Regression (LR)
b. 2 layer Perceptron with linear transfer function (PERC);
c. Decision Tree (CART);
d. Multilayer Back Propagation with one layer of 48 nodes (BP);
e. Supervised Contractive Map with 3 hidden layers of 32 nodes each (SV-CM).

The last algorithm (SV-CM) is presented here for the first time, so it is the only one that needs to be explained, provided that the other algorithms are very well known.

SV-CM is a new ANN algorithm, designed by M Buscema at Semeion Research Center in 1999 and extensively used in applied research. The equations of SV-CM are presented in Appendix A.

The BP ANN has been implemented in its enhanced version that includes "self momentum", as presented in [Buscema, 1998].

A special Evolutionary Algorithm, named Input Selection (I.S.), was used before the beginning of the training phase, in order to select dynamically the input variable with the best predictive information about the target.

Table 4.4. ANNs Training parameters.

ANN Type	Number of Layers	Number of Hidden × Layers	Learning Rate	Total Number of Weights	Number of Epochs	Final RMSE
PERC	2	0	0.1	8	2700	0.06821788
BP	3	48	0.1	289	1600	0.04760927
SV-CM	5	32 × 32 × 32	0.01	2305	3800	0.04236041

I.S. is an evolutionary system that has been generally applied to the analysis of medical datasets [Buscema, 2002; Grossi and Buscema, 2007; Penco, Buscema et al., 2008; Grossi, Marmo et al., 2008].

For the present dataset, I.S. selects four variables among the original seven: Year, Latitude, Longitude and Depth. Consequently, only these four variables were used to train the ANNs, but the Perceptron, whose dynamics is linear.

The Software used for the experimentations is: Matlab [Matlab, 2005] to implement Linear Regression and CART; Supervised ANNs and Organisms [Buscema, 2009], instead, was used to implement Perceptron, Back Propagation and Supervised Contractive Map.

Some technical detail for the ANNs training is necessary (Table 4.4):

The training phase for ANNs was executed with many trials with each type of ANN. This allowed us to choose the optimal parameters for any specific ANN (number of hidden nodes, learning coefficient and number of epochs). Each training trial needs less than 30 minutes of CPU, using a Pentium Pro Computer. To optimize each one of the three ANNs, we spent less than 5 hours of CPU time.

4.5 Function Approximation: The Testing Phase and the Results

The testing set was composed of 167270 events, randomly selected from the main dataset. The SV-CM ANN performs better than the other machines, showing learning capacities from all points of view (Table 4.5):

There is a huge difference between the performances of the non linear systems (SV-CM, BP and CART) and the linear ones (Perceptron and Linear Regression). This is evidence of the extent to which the association between the independent variables and the dependent one is non linear (see Figure 4.2).

The detailed analysis of the results of the SV-CM ANN points out another interesting finding: the SV-CM estimation of the Magnitude of the 167270

Table 4.5. Testing results.

Leaving Machines	Number of Input	RMSE	Linear Corr.	Magnitudo_ERROR	Tau
SV_CM	4	0.043075	0.854701	.387299	−22368.72
BP	4	0.047574	0.819427	0.435622	−27285.08
CART	7	0.050017	0.809565	0.453704	−31781.97
Perception	7	0.068000	0.559345	0.659792	−55746.20
Linear Regression	7	0.068002	0.559328	0.660036	−179580.81

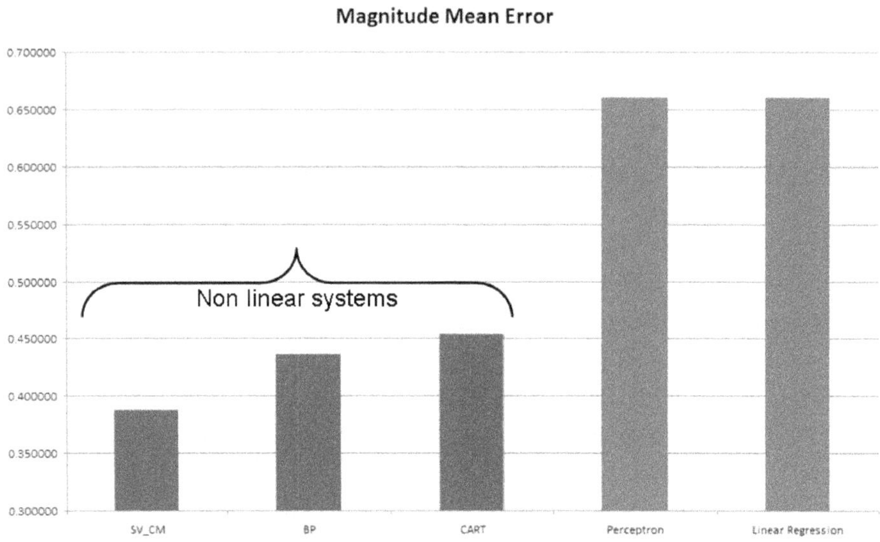

Fig. 4.2

testing events presents a very small error approximation: the absolute error is less than 0.1 of magnitude in the 18% of cases, less than 1.0 in the 94% of cases, and thus more than 1.0 only in 6% of cases (see Table 4.6).

The error approximation of this ANN grows, therefore, in quasi-log way (see Figure 4.3).

Consequently, the SV-CM function interpolation is very robust. This means also that the four inputs used as independent variables seem to carry on useful information to estimate the magnitude of the quakes. This information is highly non linear and is not time dependent. In fact, the ANN we have used to estimate the magnitude is a combinatorial system.

Finally, we can suggest that a relevant part of the magnitude of quakes is structured and cyclic. Adding to this dataset other information linked to the

Table 4.6. Distribution of Error of Magnitude in Testing (SV-CM Ann).

Magnitude error	Number of Events within the interval	% Correctness	Magnitude error	Number of Events within the interval	% Correctness
Error < 0.1	30204	0.18613420	Error < 2.8	162128	0.99912490
Error < 0.2	58578	0.36099090	Error < 2.9	162164	0.99934680
Error < 0.3	83110	0.51217110	Error < 3.0	162189	0.99950080
Error < 0.4	103166	0.63576750	Error < 3.1	162211	0.99963640
Error < 0.5	118440	0.72989460	Error < 3.2	162226	0.99972890
Error < 0.6	129950	0.80082580	Error < 3.3	162233	0.99977200
Error < 0.7	138337	0.85251120	Error < 3.4	162245	0.99984590
Error < 0.8	144393	0.88983180	Error < 3.5	162248	0.99986440
Error < 0.9	148967	0.91801940	Error < 3.6	162256	0.99991380
Error < 1.0	152267	0.93835580	Error < 3.7	162259	0.99993220
Error < 1.1	154642	0.95299190	Error < 3.8	162262	0.99995070
Error < 1.2	156469	0.96425090	Error < 3.9	162265	0.99996920
Error < 1.3	157793	0.97241020	Error < 4.0	162266	0.99997530
Error < 1.4	158787	0.97853580	Error < 4.1	162266	0.99997530
Error < 1.5	159576	0.98339800	Error < 4.2	162267	0.99998150
Error < 1.6	160148	0.98692300	Error < 4.3	162268	0.99998770
Error < 1.7	160562	0.98947440	Error < 4.4	162268	0.99998770
Error < 1.8	160898	0.99154500	Error < 4.5	162269	0.99999390
Error < 1.9	161182	0.99329510	Error < 4.6	162269	0.99999390
Error < 2.0	161411	0.99470630	Error < 4.7	162269	0.99999390
Error < 2.1	161564	0.99564920	Error < 4.8	162269	0.99999390
Error < 2.2	161719	0.99660440	Error < 4.9	162269	0.99999390
Error < 2.3	161827	0.99727000	Error < 5.0	162269	0.99999390
Error < 2.4	161919	0.99783690	Error < 5.1	162269	0.99999390
Error < 2.5	161995	0.99830530	Error < 5.2	162269	0.99999390
Error < 2.6	162049	0.99863810	Error < 5.3	162269	0.99999390
Error < 2.7	162089	0.99888460	Error < 5.4	162270	1.00000000

time arrow (geomagnetic dynamics, meteorological features, etc.), it should be possible to increase the accuracy of the estimation with important implications for civil protection purposes.

4.6 Classification Test: Recognition of Extreme Events

In the first classification experiment we assigned each event to one of three classes (multinomial classification), according to its Magnitude (see Table 4.7):

The input variables remained the seven of the original dataset: Year, Month, Day, Time, Latitude, Longitude and Depth.

Then, we divided the whole sample into two random subsets: the training set (167271 records) and the testing set (167270 records). We trained and

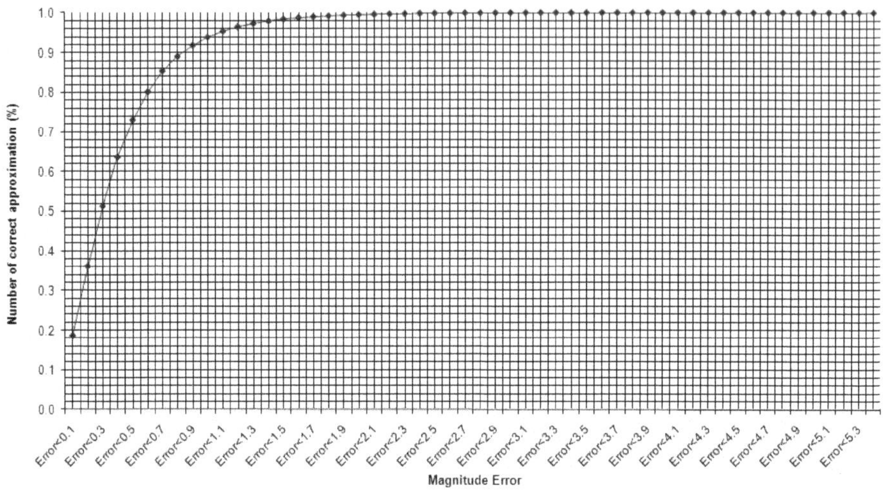

Fig. 4.3 Percent of Events within the interval.

Table 4.7. The 3 classes of Magnitude.

	Number of Events
M < 3	65993
3 <= M < 4	92311
M >= 4	166237

tested five different types of learning machines:

1. Naïve Bayes classifier (Matlab, 2005);
2. KNN classifier (Matlab, 2005);
3. SVM classifier (Matlab, 2005);
4. BP ANN (Buscema, 2009);
5. SV-CM ANN (Buscema, 2009).

Each learning machine spent, more or less, the same time for the training: about 2 hours of CPU time of a Pentium Pro computer. This is because many trials were executed to optimize the parameters of each learning machine. The best topology of SV-CM was composed of three hidden layers, of 48 nodes each and a learning rate of 0.01, while the best BP was composed of one hidden layer of 48 nodes and the same learning rate.

Table 4.8 shows the results of the testing.

Table 4.8. The testing results.

ANN	M < 3	3 ≤ M < 4	M ≥ 4	A. A.	W. A.	Errors
SV-CM	75.44%	62.27%	88.35%	75.35%	78.30%	35214
BP	66.57%	62.02%	83.72%	70.77%	74.06%	42096
SVM	69.32%	47.44%	86.95%	67.90%	72.12%	45243
Naive Bayes	79.40%	39.18%	78.43%	65.67%	67.44%	52836
KNN	66.88%	43.02%	77.13%	62.34%	65.33%	56254

Legend: A. A. = Arithmetic Accuracy; W. A. = Weighted Accuracy

Table 4.9. SV-CM Testing Results.

		SV-CM Classification		
Confusion Matrix		Class 1: M < 3	Class 2: 3 ≤ M < 4	Class 3: M ≥ 4
Real Classes	Class 1: M < 3	**75.44%**	24.15%	0.41%
	Class 2: ≤ M < 4	14.39%	**62.27%**	23.35%
	Class 3: M ≥ 4	0.89%	10.76%	**88.35%**

This time, SV-CM performed considerably better than the other classifiers. The SV-CM confusion matrix is shown in Table 4.9.

Table 4.9 has a few interesting features:

1. the SV-CM recognizes in an excellent way the big events, operates a good recognition of the small ones, but it fails to perform with the 'intermediate' events. More specifically, most of the times the SV-CM confuses the big and the middle events;
2. SV-CM attributes only a little number of big events to the small ones category, and vice versa.

This other test, in short, demonstrates, once again, the theoretical predictability of the Magnitude of the quakes even on the basis of an elementary battery of independent variables of a global dataset, without temporal information.

Furthermore, in this experiment we have also shown that big events are not less predictable than the small ones [Faenza, 2003].

The second experimentation aims to evaluate whether a learning machine is able to recognize the extreme earthquakes (over 5 of magnitude). But in the original dataset the extreme earthquakes are rare: only 41426 out of 324541 (the 12.76% of the whole sample). For this reason, we generated a training sample using only the 25% of the whole sample (83581 events), with the 50% of the earthquakes over 5 of magnitude (20713 events). This strategy aims

Table 4.10. Extreme earthquakes.

Type of ANNs	Events $<$ 5M Specificity	Events $>=$ 5M Sensitivity	Arithmetic Mean	Weighted Mean	Number of Errors
SV-CM	66.89%	83.91%	75.40%	68.14%	90091
BP	67.52%	82.38%	74.95%	68.61%	88760
SN	70.15%	79.15%	74.65%	70.81%	82524

at creating two not-too-unbalanced classes for the training of the learning machines. Consequently, we included into the testing set the other 240960 events (20713 of which are over magnitude 5).

After many trials, we have chosen three different ANNs, using 4 inputs (Year, Latitude, Longitude and Depth) and two output units (Events $<$ 5M and events $>=$ 5M):

1. BP with one layer of 32 hidden units;
2. SV-CM with three layers of 12 hidden units each one;
3. Sine Net (SN see Buscema 2006) with one layer of 48 hidden units.

The training phase was very fast: about 5 minutes of CPU time in a Pentium Pro for each ANN. Table 4.10 presents the testing results:

From a global point of view, these results show a good sensitivity and a moderate specificity.

4.7 Temporal Prediction of the Single Events

All the classifications and the estimations showed till now have executed in a rigorous blind way, but the time interval of the testing set was the same time interval of the training set. Therefore, these experimentations do not demonstrate the temporal prediction capability of the ANNs.

So, it is necessary to complete our experimentations with another task, where the last event of the training set is happened before of the first event of the testing set.

Consequently, we have divided the whole sample in two temporal halves:

a. The training set: all the seismic events from the 1st January of 1976 up to the 31st December 1993 (155730 events);
b. And the testing set: all the seismic events from the 1st January of 1994 up to the end of 2002 (168811 events).

This experimentation will verify some of the hypotheses we have formulated in this paper:

1. How each single seismic event is predictable at a specific latitude, longitude, depth and time, before it happens;
2. How much the prediction capability of the model decays along the time;
3. How to build a prediction system able to work within a specific temporal window, before to be re-trained.

Because we cope with a temporal prediction, we have used another cost function to evaluate our task: the temporal trend [Buscema and Sacco, 2000]:

$$\begin{aligned} &Legend \\ E(n) &= \text{Magnitude of Real Event at the time n;} \\ P(n) &= \text{Magnitude of Predicted Event at the time n;} \\ \Delta T(n, n+1) &= \text{Local Trend;} \\ T^* &= \text{Global Trend percent;} \end{aligned}$$

if $(E(n+1) - E(n)) \cdot (P(n+1) - P(n)) > 0$ then $\Delta T(n, n+1) = 1$ else $\Delta T(n, n+1) = 0$;

$$T^* = \frac{1}{N-1} \sum_{n=1}^{N-1} \Delta T(n, n+1); \quad \text{Null Hypothesis: } T^* = 0.5$$

The global trend percent indicates how much the model understands the seismic event temporal dynamics. The number of trend changes is particularly relevant in this test. These changes are known as "**critic points**" in financial forecasting. So, we will counts also the percent of critic points that the model is able to predict, because they represent the temporal fingerprint of the dynamical system we are studying.

To avoid possible misunderstandings of this experiment, we need to compare our model prediction capability with a "stupid" predictor, as well as in financial forecasting researchers use to do [Buscema and Sacco, 2000]. Consequently we have implemented two trivial predictors:

1. the first one for a test of prediction accuracy:

$$P(null) = \frac{1}{N-1} \sum_{n=1}^{N-1} |E(n+1) - E(n)|.$$

If the average error of the trivial predictor (P(Null)) is smaller or equal to the average error of the predictive model, then the predictive model is useless.

2. the second one to understand how much the temporal trend of the events is random:
   if $(E(n+1) - E(n)) > 0$ then $\Delta T(n, n+1) = 1$ else $\Delta T(n, n+1) = 0$;
   $T(null) = \frac{1}{N-1}\sum_{n=1}^{N-1} \Delta T(n, n+1)$; Random dynamics: $T(null) = 0.5$

More the T(*null*) predictor is bigger than 0.5, more the dynamics present a specific rule based on the **concordance** of its derivatives; more the T(*null*) is smaller than 0.5, more the dynamics presents a specific rule based on the **discordance** of its derivatives.

To implement this new experiment we have trained a SV-CM (Buscema, 2009) with seven Input (Year, Month, Day, Time, Latitude, Longitude and Depth), one Output (Magnitude) and three layers of 48 hidden nodes each one. The SV-CM was trained on the 155730 events of the training set (from 1976 to 1993) for 224 epochs (quite 1 hour of a Pentium Pro computer) and reached an RMSE = 0.04556426.

The testing was executed on 168811 events (from 1994 to 2002) and in Table 4.11 the results are segmented by year:

Some considerations are needed:

1. The prediction accuracy of the model decreases slowly with the time.

Table 4.11. Prediction Results.

Year	Prediction Mean Error	Pred(Null) Mean Error	Trend_Prediction	Trend(Null)
Year 1994	0.392269	1.115867	77.93%	29.97%
Year 1995	0.489166	1.034368	73.54%	30.13%
Year 1996	0.519026	0.878952	67.12%	29.20%
Year 1997	0.534284	0.954817	67.93%	29.72%
Year 1998	0.599928	1.013928	67.71%	30.55%
Year 1999	0.593286	1.049617	68.69%	30.16%
Year 2000	0.595394	1.062345	67.11%	30.15%
Year 2001	0.638448	1.047549	65.75%	30.30%
Year 2002	0.700586	1.062128	65.66%	30.58%
All Year	0.58890261	1.016514143	67.87%	30.13%

2. The prediction accuracy is always better than the estimations of the trivial predictor (Pred(Null)) in a very sensitive way.
3. The model is able to approximate the Magnitude of all the seismic single events in 1994 with a very small mean error.
4. The capability of the model to understand the trend of all the events of the 1994 is astonishing: quite 78%. In the stock market with a predictor like this we risk to become billionaire in 1 year. This result is surprising for another reason: it seems that the magnitude of all the seismic events, in any place of the earth, is temporally linked. This impression is supported looking the results of the trivial trend predictor (Trend(Null)): in the 70% of the cases, in each year, the magnitude follows the rule of the opposite trend. That is, if the magnitude of an event in any place of the earth at the time $(n + 1)$ will increase in relation to the magnitude of another event at the time (n), then at the time $(n + 2)$, in any other place of the earth, the next event magnitude will decrease, and vice versa. Consequently, the temporal stream of the magnitude of seismic events of the earth is far away to be a random process.

As support of these considerations, in Table 4.12 we present the capability of the model to predict the critic points of the temporal seismic dynamics:

We must consider that a random prediction about the critic points would be around 25%, because every temporal sequence of three events may have four different solutions (Table 4.13):

Consequently, the prediction capability of the model to anticipate the trend inversion of magnitude in following events in any place of the earth along the

Table 4.12. Prediction of the Critic Points.

Year	%
Year 1994	72.96%
Year 1995	66.82%
Year 1996	58.44%
Year 1997	59.35%
Year 1998	58.22%
Year 1999	59.67%
Year 2000	57.51%
Year 2001	56.31%
Year 2002	56.00%

Table 4.13. Events Magnitude.

Trend Possibilities	E(n-1) vs. E(n)	E(n) vs. E(n+1)
Trend 1	UP	UP
Trend 2	DOWN	DOWN
Trend 3	UP	DOWN
Trend 4	DOWN	UP

Table 4.14. Single event from the 1st to the 7th of January 1994.

	Number of Events	Prediction Mean Error	Pred(Null) Mean Error	Trend_ Prediction	Critic Points Prediction	Trend (Null)
From 1st to 7th Jan 1994	273	0.41	1.17	83.82%	80.72%	34.93%
Jan 1 1994	29	0.37	1.19	71.43%	63.16%	21.43%
Jan 2 1994	25	0.34	1.08	87.50%	92.86%	44.00%
Jan 3 1994	36	0.43	1.14	85.71%	75.00%	41.67%
Jan 4 1994	56	0.42	1.31	87.27%	82.86%	32.14%
Jan 5 1994	52	0.46	1.08	80.39%	72.00%	44.23%
Jan 6 1994	28	0.33	1.23	92.59%	94.44%	21.43%
Jan 7 1994	47	0.40	1.13	82.61%	85.71%	34.04%

Table 4.15. Real vs estimated results data comparison.

	Linear Correlation		Trend Correlation	
ANNs Type	Richter scale	Power 10 scale	Richter scale	Power 10 scale
Back Propagation	0.821216	0.255179	0.7551	0.7558
Sine Net	0.844258	0.229412	0.7638	0.7667
Supervised Contractive Map	0.864699	0.174954	0.7813	0.7898

1994 is surprising. And the same capability remains high up to 2002. In fact, we must remind that the ANN was trained on a subset of events from 1976 up to 1993. This means that a general system of rules, partially time independent, manages the magnitude of the sequences of seismic events all over the earth.

If we consider only the prediction of the events in the first week of January 1994 (from the 1st Jan to the 7th January 1994), the model becomes much more precise (Table 4.14):

Some considerations to the Table 4.14:

1. While the prediction error is more or less the same for one day, one week or one year (see Table 4.11), the capability of the model to anticipate the trend of magnitude of sequences of single events in non related places of the earth is increased dramatically: from

78% in 1994 to 83% in the first week of January 1994, and up to 92% in some days of January 1994.
2. The prediction error in magnitude is generally 1/3 smaller than the error estimated by a trivial predictor (Pred(Null) Mean Error).
3. The capability of the model to anticipate the trend inversion (critic points) is also increased dramatically: from 73% in the 1994 up the 80% in first week of January 1994, and up to 94% in some days of January 1994.

Final consideration: the magnitude of any seismic event in any place of the earth is scientifically predictable with an acceptable approximation. Further: there is more than evidence that the seismic activity of the earth could depend by the physiology of complex global and hidden networks, whose hidden connections should be one of main target of the next researches around earthquakes.

4.8 Temporal Prediction Based on the Moving Average

The Richter metric scales the energy of each quake with a Log10. Consequently, the error of prediction that we have shown in the previous chapters may appear optimistic and the extreme events could be underestimated. To avoid this misunderstanding, we have repeated the comparisons between the real energy and the estimated one, rescaling each value with the power of 10.

We have considered only the 10600 events occurred in 1994 over the world (according to USGS data) and three ANNs, trained from 1978 up to 1993 (155730 events):

1. a Standard Back Propagation with two hidden layers of 48 hidden units each one;
2. a Sine Net with the same topology;
3. a Supervised Contractive Map with three layers of 32 hidden units each one.

In Table 4.15 we show the results.

The transformation from the Richter metric to the Power 10 metric changes dramatically the Linear Correlation fitness, but it is indifferent to the Trend Correlation fitness. The meaning of this is clear.

The change of metric shows the impossibility of the algorithms to estimate correctly the extreme events, but the capability of the same algorithms to follow the trend remains high in both the experimentations.

These results outline a fundamental concept: the presented algorithms are able to understand the global "logic" of the quakes (so, quakes are predictable), but, because of lack of information, they are still not able to predict single event with a reasonable accuracy.

To demonstrate that our hypothesis makes sense, we re-write the real values and the estimated ones of the quakes in 1994, considering the mean of a moving window:

$$\begin{aligned}
W &= \text{Windows length;} \\
n &\in i, i + W; \\
N &= \text{Number of the Events;} \\
\forall i \mid i &\in 1, N - W; \\
\bar{r}_i &= \frac{1}{W}\sum_{n=i}^{W} r(n); \quad \text{Real Values re-writing;} \\
\bar{e}_i &= \frac{1}{W}\sum_{n=i}^{W} e(n); \quad \text{Estimated Values re-writing.}
\end{aligned}$$

The mean of the quakes in our dataset is about 50 every day. If our preprocessing window is dynamically generated as the average of 100 seismic events at the time, this means that our precision in prediction has more or less a granularity of two days. That is to say: if after this preprocessing, average based, our algorithms are able to predict extreme events (in power 10 metric) with a reasonable accuracy, consequently this means that we are able to predict extreme quakes within a temporal span of two days in a small number of selected regions of the world.

Figures 4.4–4.6 show the comparison between real seismic events and predicted one in the three algorithms trained, with a window length of 100 events at each step.

Figures 4.7–4.9 show the same process using a window average of 400. This means a precision of prediction within about 8 days.

Table 4.16 show the results of Table 4.15, updated with the new results generated using the window average.

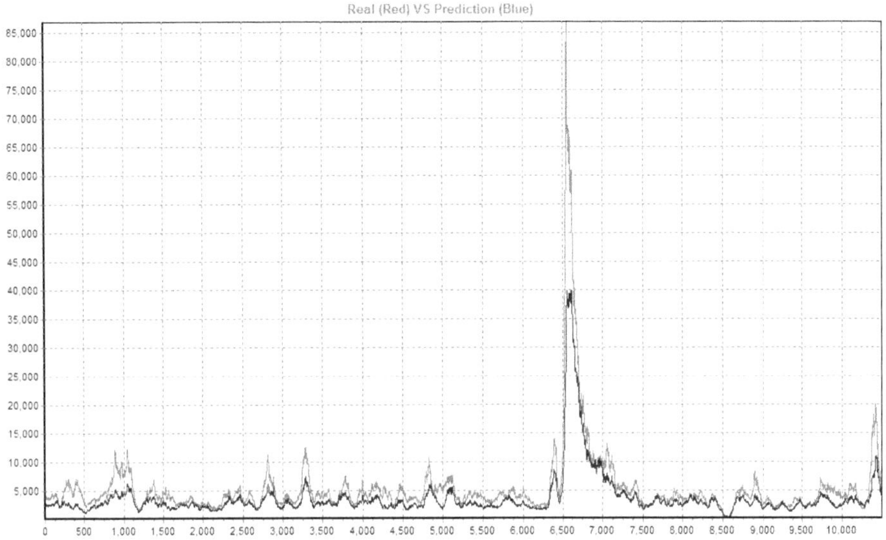

Fig. 4.4 Back Propagation — Testing 1994 — Window Average = 100.

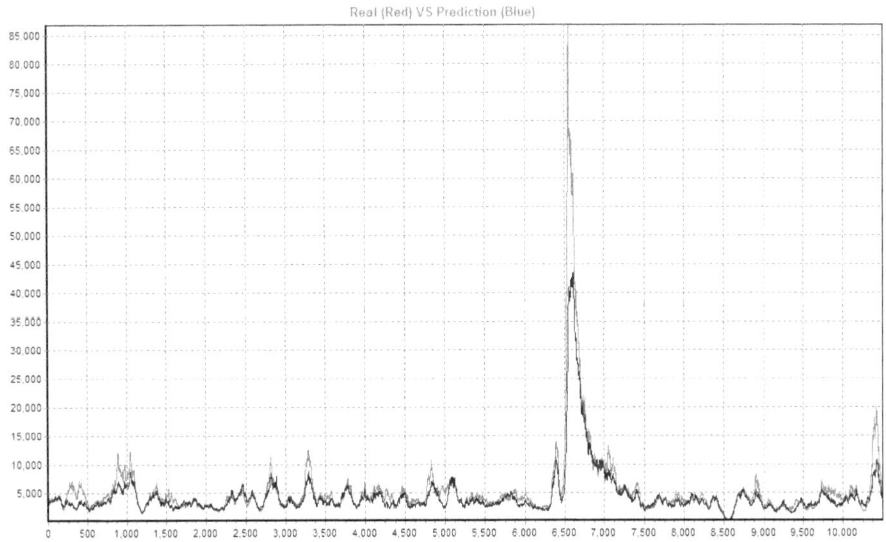

Fig. 4.5 Sine Net — Testing 1994 — Window Average = 100.

It is evident how the window average preprocessing increases dramatically the prediction capabilities of the all algorithms, in terms of real quantity of energy expressed (Linear Correlation), than in terms of temporal trends.

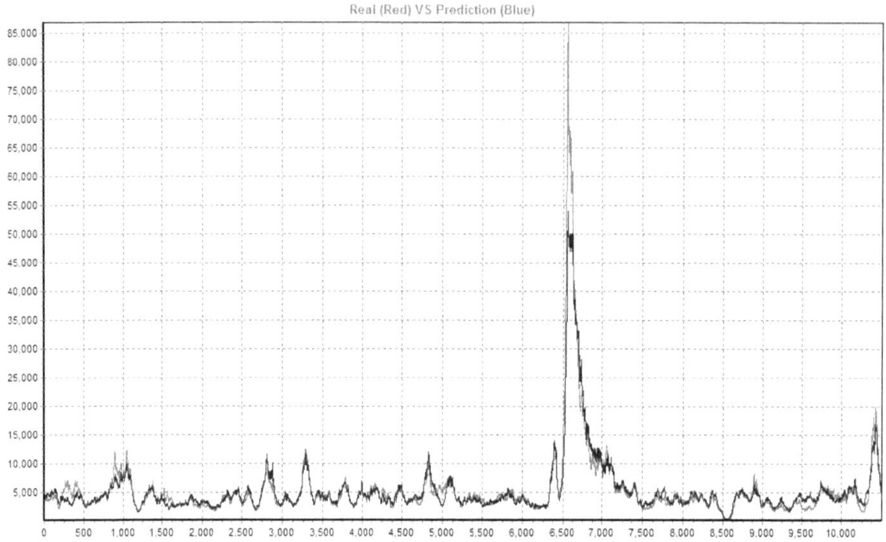

Fig. 4.6 Supervised Contractive Map — Testing 1994 — Window Average = 100.

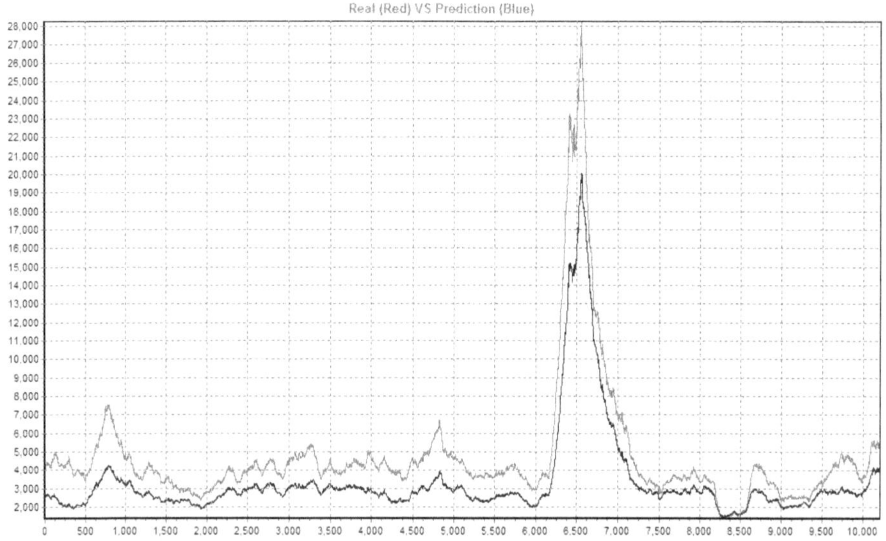

Fig. 4.7 Back Propagation — Testing 1994 — Window Average = 400.

We remind that all the algorithms were trained, with data from 1978 to 1993, using the Richter metric. Further, the testing set cover all the 10600 events recorded during 1994.

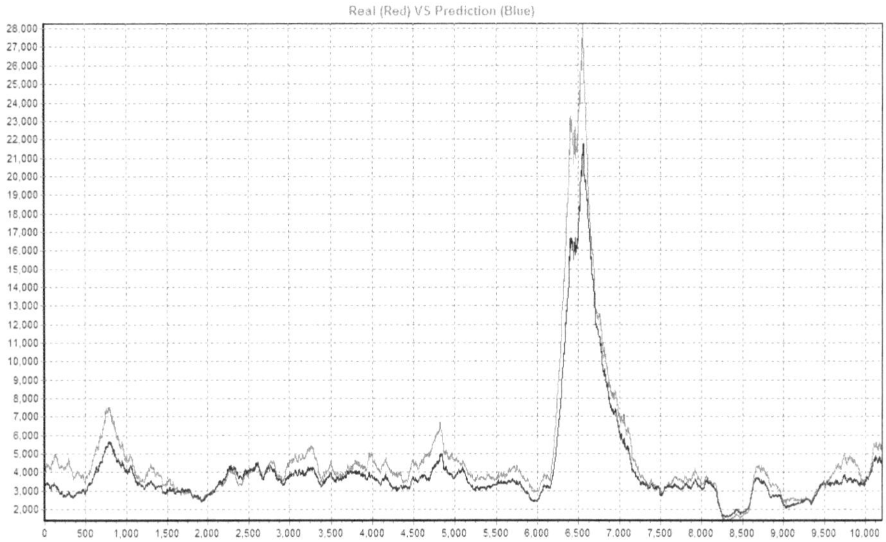

Fig. 4.8 Sine Net — Testing 1994 — Window Average = 400.

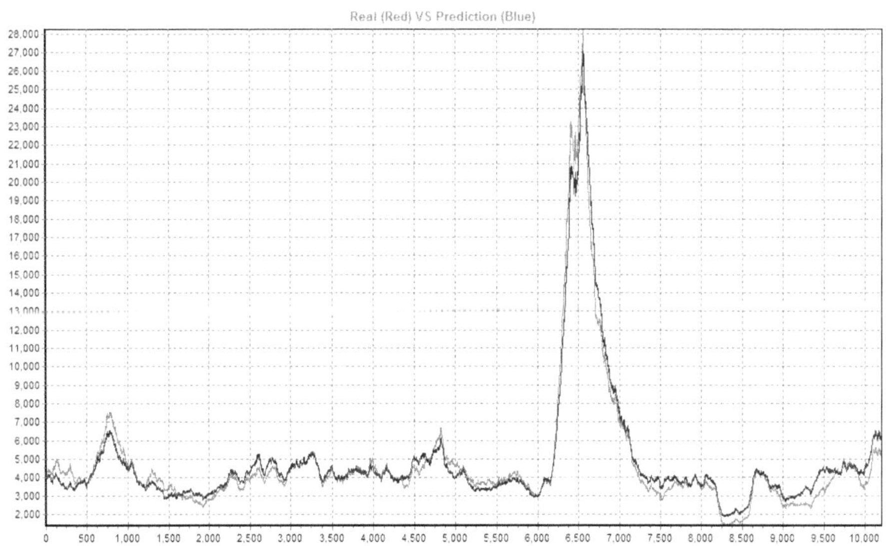

Fig. 4.9 Supervised Contractive Map — Testing 1994 — Window Average = 400.

We assume at this point demonstrated experimentally our hypothesis, that is: seismic events are not random, but they are predictable considering all the earth as a global system.

Table 4.16. Real vs. estimated results data comparison.

	Linear Correlation				Trend Correlation			
ANNs Type	Ritcher Scale	Power 10 Scale	Power 10 Scale Average Window = 100	Power 10 Scale Average Window = 400	Ritcher Scale	Power 10 Scale	Power 10 Scale Average Window = 100	Power 10 Scale Average Window = 400
Back Propagation	0.821216	0.255179	0.9655	0.9822	0.7551	0.7558	0.7835	0.7901
Sine Net	0.844258	0.229412	0.9618	0.9867	0.7638	0.7667	0.7971	0.8041
Supervised Contractive Map	0.864699	0.174954	0.9682	0.9874	0.7813	0.7898	0.8139	0.8205

We remind that we have used in this experiment trivial data. Including into the dataset new variables is the main way to reach the predictability of any single event, extreme events included.

4.9 Conclusions

The concept of predictability has been developed in the last decades as an outcome of the mathematical theory of chaotic system. Although the original concept was introduced in the framework of weather forecast, the scientific definition of predictability is based on the definition of Lyapunov exponent in dynamical system theory. The latter is a well defined mathematical procedure which can be applied, in principle, to any physical systems whose dynamics is defined in terms of a suitable set of differential equations. The above definition of predictability is not operational feasible for the case of complex systems where, in principle, the number of degrees of freedom is infinite. However, as long as the dynamics of a complex system can be represented as a set of differential equations, one can defined suitable approximations for computing the Lyapunov exponents which allow a reasonable measure of predictability. This is certainly the case for turbulent flows (a classical example of complex systems) or weather forecast.

Within the limitation of our present knowledge, one can hope to get some hints on the predictability properties of complex systems by suitable statistical analysis of experimental data. This approach is rather difficult and may lead to controversial results as soon as the statistical analysis is based on some physical assumptions on the basic law underlining the problem. A possible neutral approach is based on the use of Artificial Neural Networks (ANN) aimed at disentangling existing rules underlying the data. However, applications of ANN require at least two basic ingredients: the choice of the physical variables and a rather extensive data set.

In the last 25 years applications of ANN in forecast and hidden cast of many complex phenomena have been done in different fields and a huge literature exists on the subject. In most cases, ANN can be used as an operative tool among others to improve the overall quality of the forecast. In this paper we illustrated how the above mentioned approach can be applied to study the predictability issue of earthquake dynamic.

Our work is based on two major ideas:

— earthquake dynamics are global events on the earth
— within all possible class of ANN, we use a particular adaptive system which, following previous applications, is considered the best one for the case under consideration, although the use of standard Back Propagation does not change significantly our results.

Our results indicates that there exists a significant predictability in earthquake dynamics once we consider the space (global) and time average (few days) variable, i.e., once we limit on global quantities. Although our results are not relevant for any operational purpose, still the notion of earthquake predictability, within the limitation of our present study, is clearly demonstrated by our analysis. It is worth reminding that our original data set does not contain any physical quantities (like local stress field or geological/morphological information). We argue that a more refined data set in terms of significant physical quantities may overcome some of our present limitations. We also argue that our approach may inspire more work based on the same theoretical ideas to improve the notion of earthquake predictability and to disentangle the hard scientific issues behind such a notion.

Appendix A: Supervised Contractive Map Equations

Legend:

$$\begin{aligned} l &= \text{number or name of the ANN layer;} \\ u_i^l &= \text{values of the all } i-th \text{ nodes of the } l-th \text{ layer;} \\ w_{i,j}^l &= \text{weight matrix connecting the layer } l-1 \text{ to the layer } l; \\ C^l &= \text{Number of nodes of } l-th \text{ layer;} \\ t_i &= \text{value of } i-th \text{ of the dependent variable;} \\ LCoef &= \text{ANN learning rate.} \end{aligned}$$

Signal Transfer from Input layer to Output layer:

$$CNet_i^l = \sum_j^{C^{l-1}} u_j^{l-1} \cdot \left(1 - \frac{w_{ij}^l}{C^{l-1}}\right); \tag{1}$$

$$INet_i^l = \sum_j^{C^{l-1}} u_j^{l-1} \cdot w_{ij}^l; \tag{2}$$

$$u_i^l = \sin\left(INet_i^l \cdot \left(1 - \frac{\sin(CNet_i^l)}{C^{l-1}}\right)\right). \tag{3}$$

Weights update:

$$\delta_i^{out} = \left(t_i - u_i^{out}\right) \cdot \cos\left(INet_i^{out} \cdot \left(1 - \frac{\sin(CNet_i^{out})}{C^{out-1}}\right)\right); \tag{4}$$

$$\begin{aligned}\delta_i^{hid} &= \sum_k^{Num^{hid+1}} \left(\delta_k^{hid+1} \cdot w_{ki}^{hid+1}\right) \cdot \cos \\ &\quad \times \left(INet_i^{hid} \cdot \left(1 - \frac{\sin(CNet_i^{hid})}{C^{hid-1}}\right)\right);\end{aligned} \tag{5}$$

$$\Delta w_{ij}^l = LCoef \cdot \delta_i^l \cdot u_j^{l-1} \cdot \left(1 - \frac{w_{ij}^l}{C^{l-1}}\right). \tag{6}$$

This ANNs calculates two net inputs for each node: a classic weighted input (Equation 1) and a contractive input (Equation 2). This second net input tends to decay or to increase when the positive or negative value of the weight (w) becomes close to a specific constant (C).

Equation 3 activates the each node according to a sine function of the two net inputs (the contractive input works as a harmonic modulation of the weighted input). The vantages and the disadvantages of sine transfer function to work properly into the topology of Multilayer Perceptron was already analyzed in the scientific literature [Le Cun, 1991, 1998].

Equation 4 shows a typical error calculation using the distance between the desiderate output and the estimated output, times the first derivative of sine transfer function.

Equation 5 works in the same way of Equation 4, but using the chain rule to calculate the local error of each hidden unit.

Equation 6 updates the weight matrices, using a typical back error propagation, with a contractive factor useful to limit a extreme growing of each weight value.

References

[1] Y. Y. Kagan, "Are earthquakes predictable?" Special Section-Assessment of schemes for earthquake prediction, *Geophysics. J. Int.* (1997) 131, 505–525.

[2] Y. Y. Kagan and D. D. Jackson, "Probabilistic forecasting of earthquakes," *Geophysics. J. Int.* (2000) 143, 438–453.

[3] H. Kanamori, "Earthquake prediction: An overview, International Handbook of Earthquake and Engineering Seismology," Vol. 81B, 2003.

[4] M. L. Sharma and M. K. Arora, "Prediction of seismicity cycles in the himalayaa using Artificial Neural Network," *Acta Geophysica Polonica*, vol. 53, no. 3, pp. 299–309.

[5] A. Panakkat and Adeli "Neural Network models for earthquake magnitude prediction using multiple seismicity indicators" *Int. J. Neural Systems*, vol. 17, no. 1, 2007, 13:33.

[6] Suratgar, Setoudeh, and Salemi, "Magnitude of Earthquake Prediction Using Neural Network" in Fourth International Conference on Natural Computation, DOI 10.1109/ICNC.2008.781, IEEE 2008.

[7] E. Grossi and M. Buscema, "Introduction to artificial neural networks" 2007, *European J. Gastroenterology and Hepatology* 19:1046–1054, 2007.

[8] S. Penco, M. Buscema, M. C. Patrosso, A. Marocchi, and E. Grossi, "New application of intelligent agents in sporadic Amyotrophic Lateral Sclerosis identifies unexpected specific genetic background," *BMC Bioinformatics* 9:254, 2008.

[9] E. Grossi, R. Marmo, M. Intraligi, and M. Buscema, "Artificial Neural Networks for Early Prediction of Mortality in Patients with Non Variceal Upper GI Bleeding (UGIB)," *Medical Informatics Insight* 1:1–13, 2008.

[10] M. Buscema, "Back Propagation Neural Networks" Substance Use and Misuse, 33(2), 233–270, 1998.

[11] M. Buscema, "Supervised ANNs and organisms" ver. 15.0, Semeion Software # 12, Rome, Italy, 1999–2009.

[12] M. Buscema, "Input Selection (I.S.)" ver. 2.0, Semeion Software #17, Rome, Italy, 2002.

[13] MATLAB, "The language of technical computing" ver. 7.1., MathWorks Inc.; 1984–2005.

[14] L. Faenza, W. Marzocchi, and E. Boschi, "A non-parametric hazard model to characterize the spatio-temporal occurrence of large earthquakes; an application to the Italian catalogue" in *Geophys. J. Int.* (2003) 155, 521–531.

[15] M. Buscema, S. Terzi, and M. Breda, "Using a sinusoidal modulated weights to improve feed forward neural network performances in classification and functional approximation problems" WSEAS TRANS on Information Science and Applications, Issue 5, vol. 3, May 2006.

[16] M. Buscema and P. L. Sacco, "Feed forward networks on financial predictions: the future that modifies the present" in *Expert Systems*, July 2000, vol. 17, no. 3, 149–170.

[17] Y. Le Cun, L. Kanter and S. A. Solla, "Second Order Properties of Error Surfa Learning Time and Generalization", Advances in Neural Information Processing Systems, vol. 3, pp. 918–924. San Mateo, CA, Morgan Kauffmann, 1991.

[18] Le Cun et al., "Efficient Backpropagation Neural Networks: Tricks of the Trade" vol. 1524 of Lecture Notes in Computer Science, Chapter 1, pp. 9–50, Springer, 1998.

5

Swarm Sensor Networks

Francesco Fedi*, Sabino Cacucci*, Simone Barbera†,
Ernestina Cianca†, Marina Ruggieri†, and Cosimo Stallo†

**Space Software Italia*
†*University of Roma Tor Vergata — CTIF_Italy*

5.1 Introduction

Sensor networks consist of spatially distributed autonomous sensors to cooperatively monitor physical or environmental conditions, such as temperatures, sound, vibration, pressure, motion, pollutants, etc. [1, 2]. Various sensor networks are now being used in many industrial and civilian applications including environmental and inhabitant monitoring, surveillance for safety and security, automated health care, intelligent building control, object tracking, traffic control, etc. Sensor networks are made up of multiple independent sensors. Each sensor is usually equipped with a small-sized microcontroller, a radio transceiver or other wireless communication device and an energy source. Each sensor device or sensor node can be treated as a mobile sensor or robot. Limited bandwidth, constrained battery power, topological limitation and reduced transmission power represent strict constraints that have to be taken into account in the system design.

Most sensor applications are domain-specific and impose different constraints on the design of sensor nodes.

General requirements of wireless sensor networks are:

- ☐ energy efficiency: since it may be not practical to re-charge sensor nodes in the field due to cost or environmental restraints, energy represents one of major resource placing constraints on development and deployment of sensor networks;

- ☐ cheap;
- ☐ ad-hoc routing capability: Some sensors might not be in the reach of a base station, but might need to hop via other sensor nodes acting as relay devices between sensor nodes and base station or other nodes;
- ☐ easy to deploy: in many cases, a structured deployment of sensor nodes might be not possible and self-organisation of sensors becomes an important requirement.

Considering the above requirements, the design and implementation of Service Discovery and Routing schemes that are able to effectively support information exchange and processing in Wireless Sensor Networks (WSNs) can be a challenging task.

In sensor networks the classical client/server data distribution paradigm does not seem to be adequate mainly for three reasons. (1) sensor nodes communicate via unreliable wireless media and, hence, clients cannot rely on the accessibility of their servers; (2) typical request/response protocols for client/server model are built upon point-to-point message exchange while the communication in wireless sensor networks can exploit the broadcast nature of the wireless medium which inherently allows the reception of the same message to more than one neighbor; (3) it cannot be assumed for mobile sensor networks that services offered by a node remain accessible for a certain amount of time. The extension of the above listed capabilities to service discovery could result into complex protocols with bandwidth and power consumption, which is hardly acceptable for resource-constraint sensor networks. A new paradigm is emerging which could provide for data distribution and sharing. This is based upon the Swarm Intelligence (SI) paradigm from which it also inherits the name.

Firstly defined in [3] as *any attempt to design algorithms or distributed problem solving devices inspired by the collective behaviour of animal societies*. The ant/bee colony can be seen as a distributed adaptive system of smart control packets. Each of these packets makes little use of computational and energy resources to explore the network/environment. They efficiently cooperate with each other by releasing at the nodes information about discovered paths and their estimated quality. These similarities between foraging behaviours in insect societies and network routing has suggested a relatively large number of SI-based routing protocols for wired networks, satellite networks, Mobile Ad-hoc NETworks (MANETs) and, more recently, WSNs.

Some key questions to which researchers working on SI are trying to give an answer are [4, 5]:

- How collective behaviours can be controlled in a decentralized way?
- How a swarm can be self-organized, self-optimized and adapted to a evolving environment and changing needs of the swarm itself?
- How labor is distributed between members of the swarm?
- How a task can be partitioned so that it can be executed efficiently by a group of individuals, whether individuals should specialize or not and what is the best spatial organisation of the work?

This Chapter presents the basic concepts of SI paradigm and its application to sensor networks design as robust, scalable, self-organizing and self-repairing distributed systems.

5.2 Self-Organizing Network-Centric Systems

5.2.1 The Change of Paradigm: The Ecological Approach

Nowadays the concerns related to the environment gained a main priority. We have to face a global problem which relates to biosphere and whose consequence directly impacts the human genre due to the extent of their effects.

The more we study the different phenomena, the more grows our consciousness of the relationship among them, i.e., the faced problems are systemic which means they are linked and depend on each others.

We have to consider those problems as different facets of a same crisis, which is mainly a perception crisis. It is unable to cope with the complexity and interrelationship among apparently different and far problems.

Thomas Khun defined a scientific paradigm as “a constellation of conclusions-concepts, values, techniques, etc. — which are shared by a scientific community, and utilised by this community to define problems and their allowable solutions”.

It is a new vision of the problems that is based upon a deep ecological awareness, which is able to understand the key interrelationship among all of the phenomena and the fact that all of us, as individual and social being, both impact and depend on, the cyclic processes which regulate the Nature.

This vision looks at the world as a *network* of phenomena which are deeply interconnected and depend on each others.

This paradigm implies *the evolution from a hierarchical-centric view to a network-centric view of the complex systems.*

The Network-centric approach looks at the portions of a complex system, which can be systems themselves, as ecological community of entities which are related each other by a network of interdependencies links.

5.2.2 The Network-Centric Paradigm

The Network-centric approach moves the focus of the system analysis and design from the system *functions*, on which the mechanistic approach is focused, to the system organisation. Following this vision, the system is defined as an integrated whole, whose properties directly stem from the relationships among its components. Such a system is itself part of the environment where it acts and interacts with, via a feedback loop where each system action modifies the environment, whose changes impact the next actions of the system.

The system complexity still implies a hierarchy where a given system can act as a *part* of a higher level system, e.g., the environment itself. We usually address these concepts as an *organised complexity* where different levels cope with different complexity degree.

Another specific feature of the Network-centric paradigm specifically relates with the hierarchical organisation of the systems, the enaction of *system properties*, i.e., properties which emerge and then called emergent property. An *emergent property* relies upon the organisation of a given level, but cannot be derived directly from the capabilities of its components, i.e., if we decompose the system in its parts none of them exhibits (any part of) the *emergent property*. An emergent property directly stems from the interactions and the relationships among the system parts, i.e., from the network of relationships which constitutes the system organisation.

In the ecological paradigm the network is the recursive scheme to look at the complex systems which are studied as network of networked entities which, for what above reported, implies organisation of organised entities.

The network scheme supports a differentiated exchange of resources, energy, information, heat. Each exchange is supported by a proper network.

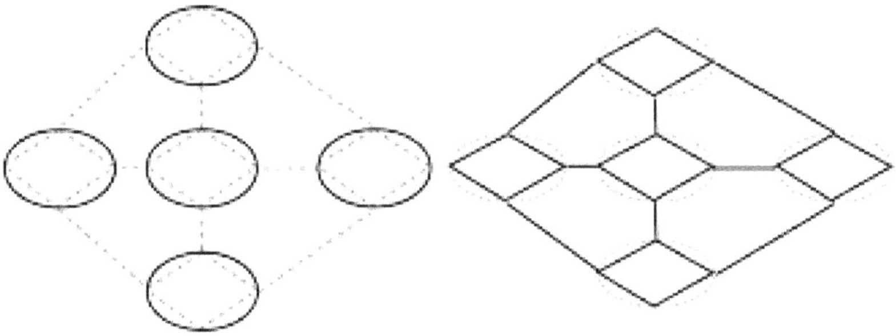

Fig. 5.1 System function view vs system organisation view.

Different network may relate each other to form *heterogeneous network of networks*. It is worth noting that the network-centric organisation does not imply a hierarchy of networks, but, a set of networks, *the systems*, which interact via networks. It is more correct to think about networks whose nodes are networks themselves.

5.2.3 The Network Scheme

The *organisational* scheme is a key concept for a systemic analysis of a system. The scheme describes a given system by defining its configuration of relationships among its components, i.e., it provides a description of the *qualitative* features of that system. There are many studies which identify the *network* as the scheme for the life; similar considerations can lead to assume this scheme to be applicable to complex systems, the living entities being a specific case of complex systems. Indeed, the network has many properties which can explain specific behaviour of complex systems.

The Feedback Loop

Each network extends in many directions as consequence circular, non linear relationships are very probable. As example, a given stimulus may travel the network till come back to its originator, so resulting into one or more feedback loops.

The *feedback* loop is a circular configuration of causally interconnected elements, where an initial stimulus propagates itself along the ring connections so that each element acts onto the next one, till the last element acts onto the source of the stimulus.

The consequence of this configuration of relationships (scheme) is that the first connection (input) is subjected to the effects of the last one (output), which results into self-control capabilities of the whole system.

An extended interpretation of the feedback scheme can be applied to information transfer within a given process and/or system.

Coming back to the network scheme is a set of potential feedback loops relationship that typically acts as a regulator, and then we can look at a network as a self-regulating, self-organising system.

Key System Properties of Network-centric Paradigm

The Network-centric paradigm relies on the following key properties of Complexity [6]:

- non-linear interaction: this can cause the rise to surprising and non-intuitive behaviour on the basis of simple local co-evolution;
- decentralized control: the natural systems are not controlled centrally;
- self-organisation: natural systems can evolve over time to an attractor corresponding to a special state of the system, without the need for guidance from outside the system. This key property is detailed in the next paragraph;
- non-equilibrium order: the natural systems, by their nature, proceed far from equilibrium. Correlation of local effects is key;
- adaptation: the dynamic systems have to be continually adapt and co-evolve in a changing environment;
- collective dynamics: the ability of elements to locally influence each other, and for these effects to ripple through the system, permits continual feedback among the evolving states of the system elements.

Another important characteristic of complex systems is their sensitivity to even small perturbations. They adapt and change their properties fundamentally as a result of the intrinsic dynamics of the system influenced by its surroundings. The application of the Complex Systems Theory in challenging sensor environments will allow developing techniques that are able to extract hypotheses about interaction networks. They can be further adopted in order to control the stressful conditions in these networks.

5.2.4 Self-Organisation

The Self-Organising Network-centric is the architectural paradigm adopted as reference model for the design of Swarm Sensor Networks.

As distributed systems tend to grow in number of components and in their geographical dispersion, deployment and management are becoming challenging. Traditional techniques for realizing distributed applications, based on client-server architecture, are showing their limits due to growing request of global knowledge and the difficulty to adapt to changing environmental conditions [7].

A new paradigm is required to manage this complexity explosion through realization of robust, scalable, self-organizing and self-repairing distributed systems [7].

A self-organising system can be defined as *a system that modifies its basic structure as a function of its experience and environment* [7].

The essence of self-organisation is that the system structure often appears with no explicit pressure or involvement from outside the system. The constraints on system structure are only internal to it and result from interactions among the system components [8]. The organisation can evolve either in time or space, can maintain a stable form or can show transient phenomena. However, the order of a self-organized system cannot be imposed by special conditions: the self-organizing behaviour is the spontaneous formation of well-organised structures or patterns from random initial conditions.

The concept of self-organisation is derived by a wide range of pattern formation processes. Biology, chemistry, geology and sociology are some areas where self-organising systems are often encountered.

Self-organisation has three important characteristics [8]. Firstly, a self-organizing system can perform complex tasks by integrating simple individual behaviours of its components. Secondly, a self-organising system is adaptable: an environmental change may influence it to generate a different task, without modifying the behaviour of its constituents. Finally, the collective behaviour of the system depends on any small difference in the individual behaviour of its components.

The results of self-organisation are global in nature as demonstrated by social insect colonies [8]. It relies on four concepts: positive feedback, negative feedback, fluctuations amplification and multiple interactions. Positive feedbacks, including reinforcement and recruitment, allow the colony to select an optimal policy for the goal. Negative feedbacks instead stabilize the collective pattern with saturation, exhaustion or competition. Random fluctuations are fundamental to enable the discovery of new solutions and can act as seeds from which structures nucleate and grow. Finally, the self-organising system relays on multiple interactions among agents, where each individual should be able to use results of the activities of other components.

5.3 Swarm Intelligence

5.3.1 The Biological Model

Biological phenomena can often be understood using physical explanations, and many systems, innate and living, share the same physical principles. It has been through these similarities and the wide applicability of the mathematical rules governing diverse behaviours that the field of "Swarm Intelligence" came to life.

A biological overview can start by dealing with the *recruitment* mechanism. Social insects need a sort of recruitment; this is defined as the mechanism increasing the number of individuals in a certain place. It can be divided into two main classes: *direct* and *indirect* mechanisms. Mass recruitment via a chemical trail is a good example of indirect recruitment. The recruiter and recruited are not physically in contact with each other; communication is instead via modulation of the environment: the trail. The recruiter deposits a pheromone on the way back from a profitable food source and recruits simply follow that trail. In a way such a recruitment mechanism is comparable to broadcasting: simply spit out the information without controlling who receives it. The other extreme is transferring information, figuratively speaking, mouth to mouth: direct recruitment. The best-known example of such a recruitment mechanism is the honeybees dance language. Successful foragers, the recruiters, perform a sort of dance which encodes information about the direction and distance of the food source found and up to seven dance followers, potential recruits, are able to extract this information [9].

Experiments implemented on some ants have shown that they modulate the measure of deposited pheromone depending on the quality of the food lineage, such that the advanced the quality, the more pheromone is withdrew and the more probable other ants are to pursue the trail to the best food source.

As on one side it is very amazing as a bug's colony collective nourishment assemblage is, even more astonishing is that the identical connection means are often utilised to accomplish a very distinct goal: the assortment of a new nest. A colony desires to choose a new dwelling under two conditions. Either the entire colony desires to proceed after the vintage nest has been decimated, or part of the colony needs a new nest location in the case of reproductive swarming, where the initial colony has developed so much that part of it is dispatched off with one or more new rulers to start a new colony. This entails that colonies require addressing inquiries very alike to the inquiries we do when going in new homes. What alternate new dwellings are available? House searching by communal bugs is even more piquant, as it is absolutely crucial for the colony that the conclusion be unanimous. Indecisiveness and contradiction are fatal. House searching has been revised in minutia in two species of communal bug only, in the ant and the honeybee.

In numerous animal species, individuals proceed in assemblies as they present cyclic migrations, journey to nourishment sources and come back to protected refuges, often over substantial distances. The action of these assemblies is routinely self-organized, originating from localized interactions between individuals other than from a hierarchical order centre. Self-organized assembly action is not peculiar of assemblies of simple insects, but may even encompass intelligent species like us. One of the most catastrophic demonstrations of collective human assembly action is gathering escaping induced by fright.

There are two farthest modes in which assemblies can decide on a main heading of movement. Either all individuals inside the assembly assist to an agreement, or a few individuals (*leaders*) have data about the group destination and direct the uninformed majority. Thus, in some species, all individuals inside an assembly share a genetically very resolute propensity to journey in a certain main heading or all are engaged in selecting a specific journey direction. Similarly, some individuals (approx. 5%) inside a honeybee swarm can direct the assembly to a new nest site. When leaders are present, the inquiry arises as to how these acquainted individuals move directional data to the uninformed majority. Similarly, in the nonattendance of leaders, how is an agreement about journey direction? Such inquiries are nearly unrealistic to address without having first evolved a theoretical structure that discovers likely mechanisms. Recently, two theoretical investigations have addressed the topic of data move from acquainted to uninformed assembly members. In 2005 it has been modelled something in which the acquainted individuals make their occurrence renowned by going at a higher pace than the mean assembly constituent in the main heading of travel. Guidance of the assembly is accomplished by uninformed individuals aligning their main heading of action with that of their neighbours. Because the acquainted individuals primarily proceed much quicker, they have a bigger leverage on the directional action of the uninformed individuals, thereby guiding the group. A second form displays that the action of a assembly can be directed by a few of acquainted individuals without these individuals supplying explicit guidance pointers and even without any individual in the group knowing which individuals own data about journey direction. Only the acquainted constituents of the assembly have a favoured main heading, and it is their inclination to proceed in this main heading that guides the group. The major distinction between the two forms lies in the occurrence or nonattendance of pointers from the acquainted individuals to the uninformed majority. In some case, leaders apparently make their occurrence renowned, while in other cases authority can originate easily as a function of data distinction between acquainted and uninformed individuals, without the uninformed individuals being adept to notify which ones have more information. It appears probable that the accurate guidance means is species-dependent. When the assembly desires to proceed very fast, for demonstration a swarm of honeybees that will not run the risk of mislaying its ruler throughout air journey, the occurrence of leaders that apparently pointer their occurrence might be absolutely crucial, as the assembly else takes a long time to start going into the favoured direction. However, when the pace of action is less significant than assembly cohesion, for demonstration because being in an assembly decreases the possibilities of predation, leaders don't require to pointer their presence. If there are no leaders, the absolutely crucial first step before an assembly can start to proceed cohesively is some grade of agreement amidst the individuals in their alignment. How this is accomplished when there are no leaders? Most probable there are a smallest number of individuals that require to be aligned in the identical main heading before the assembly can start to proceed in a specific main heading without shattering up. If the number of identically aligned individuals is underneath this threshold, the assembly does not proceed cohesively. As shortly as this threshold is passed, coordinated action is achieved. Such a non-linear transition at a threshold is renowned in theoretical physics and numbers as a stage transition. Theoretical physicists have evolved a suite of forms, termed self-propelled particle (SPP) forms, which try to arrest stage transitions in collective behaviour. SPP forms objective to interpret the intrinsic dynamics of large assemblies of individuals.

5.3.2 Stigmergy

The ants, in order to guide other ants to a food source, use pheromone trails; this process is known as *Stigmergy*.

The practical application of the stigmergy is the so-called "Ant Colony Optimization" (ACO), an algorithm inspired by the foraging behaviour of the colonies of ants. The ants have an indirect communication through the use of pheromone trails, enabling them to find the shortest path between the nest and the food source. The real application of such algorithm aims to solve discrete optimization problems.

When searching for food, the ants explore an area around the nest, randomly. When an ant finds a source, it releases a pheromone trail with concentrations depending on the quantity and the quality of the food. Additionally, the ant brings back to the nest a sample of the food. This method helps them finding the shortest path because the pheromone trail on the shortest path receives, in probability, a stronger reinforcement, since more and more ants move on it, and the pheromone trail on some longer path tends to evaporate since the ants move on the shortest.

Let's start with modeling this kind of process; we can consider a graph, called G = (V, E), where V has two nodes, called VN (Nest) and VF (Food), and E has two links, named E1 and E2, to connect VN and VF. We guess that E1 is the shortest path. About the pheromone, the model is such that we assign a value τ_i indicating the pheromone strength on the corresponding link. The probability of the ant to move through the ith path is:

$$P_i = \tau_i/(\tau_1 + \tau_2) \tag{5.1}$$

When coming back from VF to VN, the ant modifies the pheromone strength as:

$$\tau_i \text{ becomes } \tau_i + (Q/l_i) \tag{5.2}$$

The value Q is a positive feedback constant, and l_i represents the length of the path. Finally, the evaporation over time of the pheromone is an opposite

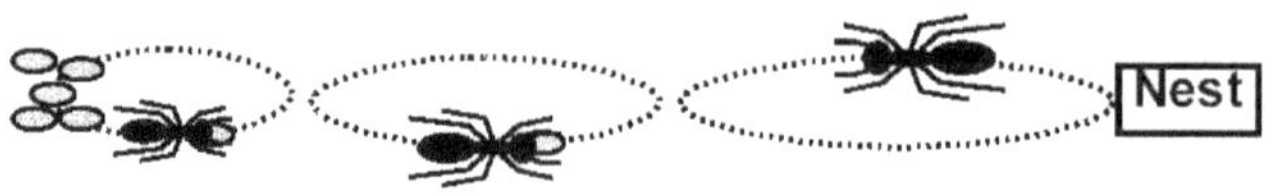

Fig. 5.2 Messor Barbarus labour division.

process that can be described through the following:

$$\tau_i \text{ becomes } (1 - \rho)\tau_i \tag{5.3}$$

In this case ρ is a parameter, €(0,1], regulating the evaporation process.

5.3.3 Labour Division

The concept of labour division is a possible example of the use of the swarm intelligence. A relevant study comes from Wilson [10]. In his study, workers in ants' colonies can be divided into two groups: small minors and larger majors. The minors accomplish daily simple tasks, while the majors do seed milling, store food, do defense tasks. Through the experimental reduction of the minors' number, Wilson noticed that some of the majors changed their tasks moving towards the tasks of the minors. This concept has been later modeled in [11] and [12], introducing several threshold values, for each kind of task; such a threshold can be considered the degree of specialization for that task. Each task emits a stimulus to attract the attention of some individual, and the task can be assigned or not to the individual depending on the stimulus and on the threshold.

Another interesting example is the application of the *Messor Barbarus* ants work. These ants, in southern Spain, retrieve seeds from a source in a bucket brigade of up to six workers. The first and smallest ant collects a seed from a source and starts to carry it along a trail towards the nest until it meets a larger worker. This larger worker takes the seed from the ant and continues to transport the seed towards the nest while the smaller ant turns and walks back towards the seed source. In turn, the larger ant continues along the trail until its seed is taken away by an empty-handed and even larger ant, and so on.

5.4 Swarm Sensor Networks

Recently, researchers have focused their attention on the application of the Complex Systems theory to the design of robust and adaptive network information systems that are self-organising and self-repairing. The network-centric paradigm considers the network as a complex system where the global behaviour is adaptive and can cope with arbitrary initial conditions, unforeseen scenarios, variations in the environment or presence of deviant agents. This leads to a radical shift from traditional algorithmic techniques to those based on achieving the desired system properties as a result of emergent behaviour, that often involves evolution, adaptation, or learning.

The swarm sensor networks represent a novel approach to the coordination of large numbers of sensors/robots whose main inspiration arises from the observation of social insects. These insects, such as ants, wasps and termites, are known to coordinate their behaviours to accomplish tasks that are beyond the capabilities of a single individual. The emergence of such synchronized behaviour at the system level is rather impressive for researchers working on multi-robot systems, since it emerges despite the individuals being relatively incapable, despite the lack of centralized coordination and despite the simplicity of interactions [13, 14].

The system-level procedure of a swarm sensor/robotic system should exhibit three functional properties that come from natural swarms and are desirable properties of such systems:

- **Robustness.** The system should be able to operate despite disturbances from the environment or the malfunction of its individuals. The loss of an individual must be immediately compensated by another one. Moreover, the coordination process is decentralized and therefore the destruction of a particular part of the swarm is unlikely to stop the whole operation.
- **Flexibility.** The swarm agents should be able to coordinate their behaviours to perform tasks of different nature.
- **Scalability.** The swarm should be able to work under a wide range of group sizes and support a large number of elements without considerably impacting the performance.

We can consider a certain number of coordination mechanisms; once again, they take inspiration from the nature. Two of the main are:

- **Self-Organisation.** In real systems, made up of real nodes (i.e., robots, sensors), the concept of Self-Organization is crucial, since it is the basic paradigm behind the autonomous behaviour of the agents; thanks to this concept, a swarm system does not need a central station driving the whole system, but is able to configure itself. On one side, in biology, the Self-Organisation is defined as the process in which global patterns of a system emerge solely from numerous interactions among the lower level components of the system itself. This is common in natural systems. Studies of self-organisation in natural systems show that interplay of positive and negative feedback of local interactions among the individuals is essential. The positive feedback cycle is balanced by a negative feedback mechanism, which typically arises from a reduction of the physical resources. In addition to these mechanisms, self-organisation also depends on the existence of randomness and multiple interactions within the system. The self-organisation models of social insects and animals have already been used as inspiration sources. On the other side, the swarm sensor networks can be considered as the engineering utilization of the self-organisation paradigm. In fact, in real networks, the self-organisation can be considered as the capability of a system to change its organisation in case of environmental change without explicit external command.

- **Stigmergy.** Previously, we emphasized the biological definition of stigmergy, defining it as the process, typical of the ants, to release pheromone trails in order to drive other ants to a food source. In real applications, when real agents, as robots, are involved, this behaviour can be seen as the property of the agent/node to modify the environment in which is moving, in order to reach a specific goal.

Below, we now focus on some problems typically encountered in the swarm sensor networks:

- **Aggregation.** Self-organized aggregation, the grouping of individuals of a swarm into a cluster without using any environmental clues, is a common behaviour observed in organisms ranging from bacteria to social insects and mammals. In swarm robotic systems it can be considered as one of the fundamental behaviours that can act as a precursor to other behaviours such as flocking and self-assembly.

- **Dispersion.** Self-organized dispersion can be advised as the opposite of aggregation and is of concern in surveillance scenarios. In this problem the challenge is to obtain uniform spreading of a swarm of robots in a space, maximizing the area covered yet remaining connected through some sort of connection channel.
- **Foraging.** This problem is motivated by the behaviour of ants which search for food sources distributed around their nest. In this problem, the challenge is to find the optimum seeking schemes maximizing the ratio of returned food to the assets committed in an environment.
- **Connected Movement.** This problem can be described as follows: how can a swarm of mobile robots, physically connected to each other, coordinate their movement such that the group moves smoothly in an environment and avoids environmental obstacles, such as holes, in a coordinated way? In the implemented studies, evolutionary approaches were used to evolve behaviours that can control a number of connected robots to avoid holes within the environment.
- **Cooperative Transport.** Ants are renowned to transport large preys to their nest through coordinating their impelling and dragging actions. Such coordination proficiency is conspicuously precious for swarm robotic schemes since it allows persons to connect forces, developing a blended force large enough to drag a hefty object. This difficulty is partially associated to the connected movement, with the distinction that it encompasses a passive object that needs to be transported.
- **Pattern Formation.** This concerns the problem of how a desired geometrical pattern can be obtained and maintained by a swarm of robots without any centralized coordination. Pattern formation may refer either to geometric or to functional pattern formation. In geometric pattern formation, the challenge is to develop behaviours such that individuals of a swarm form a desired geometrical pattern, similar to the formation of crystals. In this task, the environment is assumed to be uniform and the focus is on the use of inter-robot

interactions to create such patterns. In functional pattern formation, the pattern to be formed is dictated by the environment.

5.5 Swarm of Satellite Sensors

5.5.1 Introduction

Space engineers are currently developing autonomous systems and are envisaging space missions that would certainly benefit from a deeper understanding of the collective behaviour of similar and dissimilar agents. The possibility to have spacecraft swarms with autonomous decision capabilities would allow the realization of multi-robot planetary exploration, on-orbit self-assembly, and satellite swarms for coordinated observations. It would be possible to build large solar panels or large antennas by exploiting collective emerging behaviours. The idea of a number of satellites flying in formation has been used in a number of missions for applications ranging from X-Ray astronomy (XEUS), differential measurements of the geomagnetic field (CLUSTER II), space interferometry, the search for exoplanets (DARWIN) and others.

5.5.2 Swarm of Satellites: Theory and Architecture

Swarm intelligence methods would represent an attractive design option allowing, for example, achieving autonomous operations of formations. Simple agents interacting locally could be considered as a resource, rather than as an overhead. At the same time one would be able to engineer systems that are robust, autonomous, adaptable, distributed and inherently redundant. Besides, swarms allow for mass production of single components, thus promising for mission cost reduction. These motivations led recently a number of researchers to simulate some degree of swarm intelligence in a number of space systems and to investigate their behaviour.

A possible example consists in spacecraft which have to accomplish proximity operations and have to reach, with a group of other satellites, a very tight formation, or indeed dock with those other spacecraft. A fundamental point of spacecraft swarm operations therefore involves position and velocity control.

The final position occupied by each agent should be chosen in an autonomous way between all the possible ones according to the initial conditions imposed. Each satellite belonging to the swarm will be then able to autonomously decide what will be its final position in the target configuration, exchanging a minimum amount of information with the other swarm components.

Many researchers have faced the question whether it is possible or not to design systems in which clusters of vehicles autonomously behave in a coordinated manner performing high level tasks. In all such works the navigation technique is always the result of a trade-off between autonomy and optimality. The more the swarm of satellites is required to perform optimal manoeuvres (i.e., from the point of view of fuel consumption), the higher is the computational load for the on board computer.

5.5.2.1 Equilibrium shaping

A behaviour based navigation technique has been introduced by Izzo and Pettazzi [15, 16]; this technique is able to make a group of identical spacecraft acquire a given configuration. This navigation technique, called Equilibrium Shaping (ES), allows dealing with large swarms of spacecraft and has shown to be simple, but not optimal from the fuel consumption perspective.

The ES technique aims at steering a swarm of N homogeneous satellites to acquire a target formation in orbit in a way such that each spacecraft belonging to the swarm can autonomously decide which position it will take in the final configuration.

Each agent of the swarm is asked to follow a certain velocity field defined as the sum of two different contributions, both dependent on the inter-agent distance:

$$x_{ij} = x_j - x_i \tag{5.4}$$

The contributions define:

- A Linear Global Gather Behaviour.
- An Avoidance Behaviour.

The mathematical definition used by Gazi [17] for the desired velocity of the ith agent is:

$$v_{di} = -\Sigma_j x_{ij} c_i - b_i \exp(-x_{ij} x_{ij}/k_1) \tag{5.5}$$

In the formula above, c_i and b_i are coefficients whose values are determined by the formation geometry, while k_1 is a parameter determining the size of the avoid behaviour sphere of influence. This method produces a swarm in which each agent is pre-assigned to a particular place in the final formation. For a given set of positions in the space, the target assignment problem is the problem of associating every single agent belonging to the swarm with every element of the set on a one-to-one basis. Such a problem can be autonomously solved by the swarm defining the desired kinematical field according to the equilibrium shaping approach here described.

In such technique there is a dynamical system which has, as equilibrium points, all the possible agent permutations in the final target formation. The system is then used as a definition for the desired velocities. It is basically an approach that starts from the geometry of the final relative configuration and finds the necessary agent behaviours to reach that configuration.

The dynamic of the system can be considered as the sum of three different contributes:

- Gather.
- Dock.
- Avoid.

The *Gather Behavior* considers different global attractors toward the agents. Here each component of the swarm is as the others, allowing agent permutations.

The main reason of this contribute is to gather the spacecraft from wherever it is, and therefore this vector can never be neglected in the space around the target configuration.

The *Dock Behavior* considers different local attractors toward the agents. If the agent is in the neighborhood of the sink, this contribute cannot be neglected.

The main reason of this contribute is to handle the final docking procedure, and for this reason the sole local attractors are taken into account.

The *Avoid Behavior* introduces a relationship between two agents in proximity. For this reason, the velocity avoid contribute is a repulsive contribute.

The previously used expression for velocity can be therefore written as the sum of these contributes, becoming:

$$v_{di} = v_i^{Gather} + v_i^{Dock} + v_i^{Avoid} \tag{5.6}$$

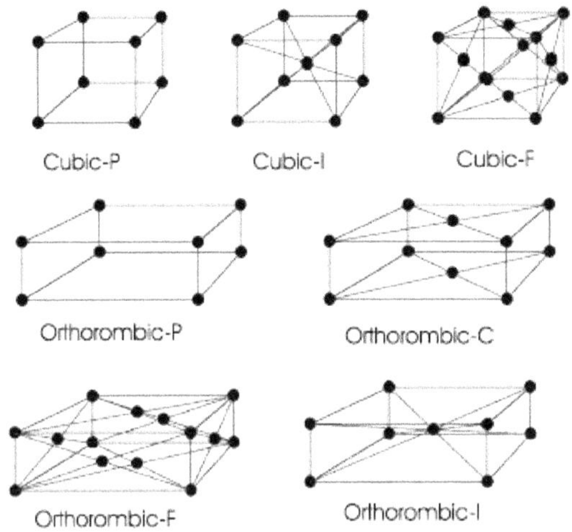

Fig. 5.3 Some possible formation of spacecrafts.

Acting on the different coefficients involved in the formulas, we can obtain several formations; typical examples are shown in the next figure [16]:

In order to reach the desired velocity, several control systems can be exploited. In particular:

- Q-Guidance [18].
- Sliding-Mode Control [19].
- Artificial Potential Approach.

The first method consists in the definition of a vector (v_g) representing the difference between the desired velocity and the actual one for an agent of the swarm. A proper feedback acts in order to bring to zero the value of such vector.

The second method, a newer method with respect to the first, aims to design a control law able to drive the system trajectory on a pre-established manifold and keep it there once obtained. The procedure can be split into two different steps: design of a sliding manifold to lead the swarm towards the desired equilibrium; derive a control law forcing the trajectories to converge to the sliding manifold.

The third method starts from the definition of an artificial potential function for the whole swarm, with minimum points at all the possible agents' permutations in the final wanted configuration. The swarm can reach the final formation, without any collisions, whenever the potential function decreases during the motion. Compared to the previous design methods, this relies upon a slightly different approach. First, a global artificial potential function is defined for the entire swarm of spacecrafts.

Next, a control law is imposed; the potential function decreases along the trajectories followed by the system. The resulting feedback cannot be obtained from the *Q-guidance* or the *Sliding-Mode Control*.

5.6 Swarm of Positioning Sensors

5.6.1 Introduction

Having *Geo-Referenced* swarm agents simplifies the organisation of the swarm towards the final formation, since the agents are aware in real-time of their position and can move according to it. A key role, in this sense, is played by the accuracy and precision of the so-called *Global Navigation Satellite Systems* (GNSS). For this reason, in this paragraph we propose an overview of the current GNSS technologies and a possible application of such technology to the swarms.

5.6.2 GNSS State of the Art

Nowadays, satellite navigation shows a very complex scenario. Global positioning and its related services and applications are no more a military asset and a strategic and predominance issue. There are several operational and proposed systems that meet the civil and/or mass market requirements. The following figure explains the current worldwide scenario:

The global navigation satellite systems distinguish between the systems already in use, the systems in the last stage and the systems in phase of study/deployment. The oldest systems are the American GPS (Global Positioning System) and the Russian GLONASS; a great difference between them, beyond the techniques and the architectures used, is given by the present efficiency, with GLONASS that is falling rapidly into disrepair, and GPS that is in full swing. Now, a special-purpose federal program has been undertaken by the Russian government; according to it, the GLONASS system has to be restored to fully deployed status by 2011.

	USA	Europe	India	Japan	China	Russia
Operational	GPS, WAAS					GLONASS
Experimental Phase or deployment phase	GPS II	EGNOS, GALILEO	GAGAN	MTSAT	BEIDOU	
Development phase or proposed	GPS III		IRNSS	QZSS	COMPASS	

Fig. 5.4 GNSS current scenario.

Concerning systems in the last stage, we can consider the so-called SBAS, *Satellite Based Augmentation Systems*, which aim to exploit the current systems improving the performances with additional satellites, ground stations and technologies; among these, the European EGNOS and the American WAAS.

In course of development, there are the systems that will aim to provide to the users the greatest possible number of services. These systems are the American GPS III and the European GALILEO. The path, to achieve the targets, consists in the exploitation of the current systems, keeping the advantages and introducing the needed improvements. Currently, tests on these systems are in progress, and the only one completely developed and in use remains GPS.

As all GNSS, GPS is made up with three segments: the space segment, that consists of the satellites needed for a global coverage, the ground segment, which consists of the ground stations for the monitoring and control, and the user segment, that consists of the receivers. Concerning in particular the space segment, the GPS constellation has 24 satellites on 6 orbital planes, all with the same inclination of 55 degrees; the orbits are circular and the name of this kind of constellation is "Walker Constellation". The height of the satellites is about 27000 km.

5.6.3 A Possible GPS Application — SoHa-Pen

The current GNSS scenario, as pointed out in the previous section, leads to the use of GPS receivers as best technology in order to locate the agent of a swarm. In particular, here we propose an application in which a GPS receiver is associated to each agent, but it is not included in the hardware implementing the swarm intelligence, but it is externally USB plugged in order to perform certain missions with a high level of security; this device is called SoHa-Pen (Software to Hardware Pen-Drive) and is here described in details.

5.6.3.1 Introduction to SoHa-Pen

When mobile physical agents, equipped with sensors and/or actuators, have to substitute for the human operator in performing tasks in risky environments, cooperative Multi-Robot/Multi-Sensor systems can be considered a meaningful example.

In this context, a device that guarantees to perform the mobile physical agents operations only when certain parameters are set has been developed. Such a device is particularly small in order to be easily integrated within the agents. The main feature is the geographical reference, which allows the use of the device to ensure a defined operational area; such a function is guaranteed thanks to a GPS receiver that communicates with a microcontroller, in which proper software handles data.

The manageability of autonomous mobile agents poses new typologies of impacts due to possible misbehaviours deriving from accidental and/or intentional faults. New impacts are possible by misbehaviour of an autonomous mobile physical agent acting in a risky environment. In order to maintain the requested safety in operations, there is the need for appropriate techniques and mechanisms to force spatial and timing mission requirements.

In order to guarantee that a device performs its own operations only in a certain area and when certain parameters are set, a portable and reconfigurable system has been conceived. The portability and reconfigurability of the system is achieved through the use of a small hardware unit combined with flexible software. The integration of a GPS receiver in the portable system can ensure that the swarm devices move in a desired time interval, in a given area at a specified velocity. The GNSS-driven system hence performs the function of a key for a fleet of devices, which can move — following their intelligent algorithm — only provided that the previously mentioned conditions are verified.

The portable system architecture can rely on a simple USB interface, and thus the system is automatically recognized by any host device. The system hence becomes like a pen-drive, whose task is to receive the mission data and to allow the device where it is applied to operate only if the geo-conditions are verified.

The pen-drive includes a microprocessor having an internal serial interface with the GNSS receiver and an external USB interface to communicate with the host computers. Thanks to this pen-drive, the operational computers are able to operate only when they receive an enabling signal. The latter is released only if the computed GNSS data are within the interval previously defined in the configuration phase. This pen-drive is useful to ensure that the fleet moves correctly; in fact, a device moving in a wrong way is excluded from the fleet, by a proper switching to a sleep state. The depicted system can be exploited in multiple applications where security is a priority and a given behaviour of the group has to be guaranteed [20, 21, 22].

5.6.3.2 SoHa-Pen architecture

The system architecture consists of three main parts:

- The *Pen-Drive.*
- The *Secondary Host PC*, inserting the mission data.
- The *Main Host PC*, controlling the mobile agent.

Configuration Phase — The preliminary step consists in introducing the mission data into the pen-drive memory. However, this is not the default operational mode. Operative Phase — By default, the pen-drive starts the comparison between the previously introduced data and the GPS computed ones.

To avoid the device starting with its default mode when we have to write the mission data into the EEPROM, a safe algorithm has been implemented. In the following figure, it is summarized the overall architecture of the system.

The microcontroller must handle three interfaces: the first one must ensure the communication with the GPS receiver, the second one with the main host PC and the third one with the secondary host PC. The GPS chipset makes available two kinds of protocols: SPI (Serial Peripheral Interface) and UART (Universal Asynchronous Receiver Transmitter). The second one has been used in order to exchange the GPS data according to the NMEA standard. On the other side, the hosts PCs interface the microcontroller through an USB serial connection.

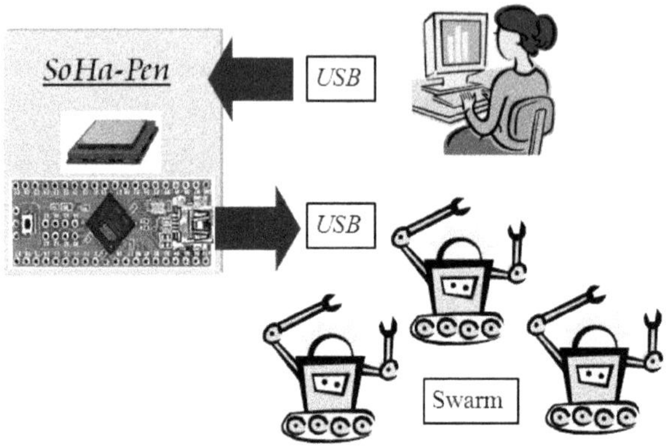

Fig. 5.5 Geo-Referenced swarm architecture.

Fig. 5.6 SoHa-Pen.

The USB serial connection allows the host PC to recognize, after installing the proper drivers, the SoHa-Pen when plugging it into the connector. Therefore, the usage is really simple. What is complicated, due to the implemented procedure, is to change the mission data and to cheat the system by a remote attack.

The math behind the algorithm is not particularly relevant; the most important parts of the embedded software concern the implementation of the protocols, both the known ones (USB and UART on the physical layer) and the custom ones (the interface with the hosts on the MAC layer). However, the "in-area function" has been written, in order to verify if a computed point is included into the mission area. This function is based on the fact that, at each iteration, it is compared the computed position ($x_{\text{comp}}, y_{\text{comp}}$) with the defined area; such area is built as a closed convex surface, in which the points are inserted clockwise. The mathematical area function is based on the fact that, once defined various points, starting from the first inserted point we consider the straight line that passes through that point and through the following one, clockwise. If the test point is "inner" with respect to that line (according to the formula below) the test is positive. This must be valid for all the points of the area (that is, for every straight line):

$$\frac{(x_{\text{comp}} - x_i)}{(y_{\text{comp}} - y_i)} > \frac{(x_{i+1} - x_i)}{(y_{i+1} - y_i)} \tag{5.7}$$

A test campaign has been done in Taranto, exploiting the SSI (Space Software Italia) swarm agents. The precision of the system can be considered the maximum achievable, according to the GPS limits, and the main limit is the indoor use of the whole architecture, being the GPS a satellite system.

References

[1] S. Hadim and M. Mohamed, "Middleware: Middleware Challenges and Approaches for Wireless Sensor Networks," *IEEE Distributed Systems Online*, vol. 7, no. 3, March 2006.

[2] C. Raghavendra, K. Krishna, and T. Znati, "Wireless Sensor Networks," *Springer-Verlag*, 2004.

[3] E. Bonabeau, M. Dorigo and G. Theraulaz, "Swarm Intelligence: From Natural to Artificial Systems," *Oxford University Press*, New York, 1999.

[4] M. Dorigo and T. Stützle, *Ant Colony Optimization*, Cambridge, MA: MIT Press, 2004.

[5] A. Engelbrecht, *Fundamentals of Computational Swarm Intelligence*. Hoboken, NJ: Wiley, 2005.

[6] P. Antoniou and A. Pitsillides, "Understanding Complex Systems: A Communication Networks Perspective," Technical Report TR-07-01, Department of Computer Science, University of Cyprus, February 2007.

[7] E. Bonabeau, V. Fourcassie, and J. L. Deneubourg, "The phase ordering Kinetics of cemetery organisation in ants," *Research Report*, Santa Fe Institute, 1998.

[8] A. Montresor, H. Meling, and O. Baboglu, "Towards Self-Organizing, Self-Repairing and Resilient Large-Scale Distributed Systems," Lecture *Notes in Computer Science*, vol. 2584, 2003, pp. 119–123.

[9] B. Christian and D. Merkle, *Swarm Intelligence, Introduction and Applications*, Springer.

[10] E. O. Wilson, "The relation between caste ratios and division of labour in the ant genus phedoile," *Behavioral Ecology and Sociobiology*, 16(1): 89–98, 1984.

[11] G. Theraulaz, E. Bonabeau, and J. L. Deneubourg, "Response threshold reinforcement and division of labour in insects societies," *Proceedings Biological Sciences*, 265(1393): 327–332, 1998.

[12] E. Bonabeau, G. Theraulaz, and J. L. Deneubourg, "Fixed response thresholds and the regulation of division of labor in social societies," *Bulletin of Mathematical Biology*, 60: 753–807, 1998.

[13] M. Ayre, D. Izzo and L. Pettazzi, "Self Assembly in Space Using Behaviour Based Intelligent Components," TAROS, Towards Autonomous Robotic Systems, Imperial College, London, Sepember 2005.

[14] A. C. Clarke. Extra terrestrial relays. Wireless World, pp. 305–308, 1945.

[15] L. Pettazzi, D. Izzo and S. Theil, "Swarm navigation and reconfiguration using electrostatic forces" *Proceedings of the 7th International Conference on Dynamics and Control of Systems and Structures in Space*, pp. 257–268, 2006.

[16] D. Izzo and L. Pettazzi, "Autonomous and Distributed Motion Planning for Satellite Swarm," *Journal of Guidance, Control and Dynamics*, vol. 30, no. 2, pp. 449–459, March–April 2007.

[17] V. Gazi, "Swarm Aggregations Using Artificial Potentials and Sliding Mode Control" *Proceedings of the IEEE Conference on Decision and Control, IEEE Publications*, Piscataway, NJ, December 2003, pp. 2848–2853.

[18] R. H. Battin, "An Introduction to the Mathematics and Methods of Astrodynamics," AIAA Educational Series, AIAA, New York, 1987, pp. 23–29.

[19] R. De Carlo, S. Zak and S. Drakunov, "The Control Handbook," CRC Press/IEEE Press, Boca Raton, FL, 1996, pp. 941–951, Chapter 57.5.

[20] M. G. Zimmermann and V. M. Eguiluz, "Evolution of Cooperative Networks and the Emergence of Leadership," *7th International Conference of the Society for Computational Economics*, June 28–29, 2001.

[21] S. Barbera, C. Stallo, G. Savarese, M. Ruggieri, S. Cacucci and F. Fedi, "A Geo-Referenced Swarm Agents Enabling System for Hazardous Applications," In: *UKSim 12th International Conference on Computer Modeling and Simulation*, Cambridge (UK), March 24–26, 2010, ISBN/ISSN: 978-0-7695-4016-0.

[22] S. Barbera, C. Stallo, G. Savarese, M. Ruggieri, F. Fedi, and S. Cacucci, "A Georeferenced Swarm Agent Enabling System: theory and demo application," *Proceedings ISMS*, Liverpool, January 2010, ISBN 978-0-7695-3973-7.

6

Semantic Technology

Maurizio Mencarini

Expert System

6.1 Introduction

"Anna, please, could you look for some material about Flex, the product we introduced at the Trade Show in Milan? I mean data, analyses, documents... some time ago Sergio and Paola sent me a report about it. Talk to them about this report, because I don't have time now. Look for something on the Internet too, and while you're searching for it, check out the competitors."

This is a marketing director who is speaking to a project or production manager. He fixed an important meeting and he needs all the documents required to prepare it. The problem is finding them.

The management of knowledge is a key issue for any sector, for any objective, and for employees of all levels... even those fortunate enough to delegate all of these search activities.

Different theories have been developed about KM, Knowledge Management. The debate is interesting and expanding, due to its links to philosophy, linguistics, information technology, knowledge and language engineering, mathematics and didactics.

Thanks to linguistic technologies we've solved the primary problem, which is information access. However, we are far away from managing all the knowledge we produce and learn in everyday life because we lack the time to do it.

Yet, wouldn't it be great if we could already know the subject of a document before opening and reading it? And if we could do this not just for a single document, but for the hundreds or thousands of documents, books, articles, Word or Excel files, private e-mails, html pages or pdf… we manage everyday, even if stored on several different computers, or hidden inside company Intranets or on a website?

Semantic technology can offer us exactly this: time.

Time to intelligently manage our knowledge. Time to integrate information, data, experiences; to update and grow our intellectual capital; and to ensure we can track down the useful information in order to make the best decisions.

6.2 The Information Management Problem

The Internet advent and the large distribution of contents have created a huge amount of information at people's disposal, and the proliferation of potentially interesting data seems endless. Thanks to the Web, anyone can access countless electronic texts.

In the same way, the amount of information hidden inside company Intranets (documents, reports files, private e-mails, etc.) is raising exponentially, making the search for specific documents more and more difficult: they must be somewhere, but traditional search techniques require too much time to find them.

Current unstructured information management systems are unable to solve the problem of collecting only useful data, so users experience the following issues:

Too many answers (overload): the system is unable to classify and arrange information effectively. It gives many irrelevant answers (the dross), and contained somewhere in this information is the answer one is looking for (the gold nugget), yet there's no time to check answers one by one.

Few or no answers (underload): the system is unable to give any answer due to the classification inefficacy or poor selection power of the search engine.

The Internet and its related technologies cannot express their true potential as long as these problems remain unsolved.

6.3 Linguistic Technology

Through linguistic technology, it's possible to "understand" the language in a way that's similar to the way people do: by collecting all the structural and lexical aspects of the text in order to understand the meaning.

The result of the linguistic analysis based on the "semantic understanding of words" is a conceptual map, i.e., a structured representation of qualifying aspects of the incoming unstructured data. The structuring of the output allows the automatic processing of the most relevant elements of the text.

There are other text automatic processing systems, but they are superficial in their analysis. For example, they completely ignore the so-called stop words, i.e., words that do not have a proper meaning (like articles and prepositions), when indeed they are fundamental in forming the meaning of sentences.

On the other hand, the linguistic analysis based on semantics takes into account every single word, just as people do when they read, listen or write.

In order to better understand this feature, consider these two sentences:

(a) *People who are sick of this city.*
(b) *People who are sick in this city.*

The meaning of these two sentences is completely different, because of the prepositions. In the sentence (a) the word "sick" means disgusted or tired of— while in the sentence (b) "sick" means ill or not in good physical or mental health.

This difference is easily understood by a person reading the two sentences, but this doesn't happen with current text management programs: from their point of view the two sentences are the same.

On the contrary, the semantic analysis can detect the difference and distinguish between the two concepts, allowing a conceptual search.

Generally, linguistic technologies have an open and scalable infrastructure, which can be easily customized in order to manage any type of content: technical documentation, encyclopedic material, financial information, newspaper articles and so on.

Let's have a look at some examples in order to show how the linguistic analysis platform works and how the semantic knowledge compares to other text processing analysis approaches.

6.4 Which Features?

When we deal with a "linguistic technology" we refer to a software which contains different integrated elements that make it a unique technology worldwide.

Its main components are:

- parser: carries out morphological, grammatical and syntactical analysis of the sentence
- lexicon: recognizes words and all their meanings
- memory: keeps trace of previous analysis outcomes
- knowledge: a representation of real world knowledge
- content representation: text contents in the form of a conceptual map

Parser

The first step to take in order to understand the meaning of a sentence is to determine the grammatical role of each word. For example, in these sentences:

(a) *There are* 40 *rows in the table.*
(b) *She rows five times a week.*

The word "rows" appears with two different grammar types: in sentence (a) the word is a noun, while in sentence (b) it is a verb.

According to traditional systems the two words are the same, while the linguistic technology assigns different meanings to them.

Recognizing a word independently of its written form is equally important; nouns and verbs have several forms:

(a) *Marcello Mastroianni was the most popular Italian actor abroad.*
(b) *Today it is rare for mature actresses to play the main role in a movie.*

In the sentences above, two forms ("actor" and "actresses") expressing the same concept are used. The linguistic technologies individuate gender — masculine/feminine — and number — singular/plural — to recognize both words as forms of "acting person."

In regards to verb forms, the linguistic technologies work likewise, associating all of them to their common meanings correctly, instead of individuating *n* different words, as other systems do.

An "advanced" parser manages the grammar characteristics of the sentence completely and optimally. It carries out the logical analysis, providing a solid base for content processing.

Lexicon

The information related to all the possible meanings of words is fundamental in order to process text content with high precision.

Being unable to detect different meanings brings a misleading understanding of the phrase, consider this example:

(a) *The race car driver was injured.*
(b) *I used the long driver.*
(c) *The driver was installed in the computer.*

The word "driver" is ambiguous, because its meaning depends on context. In order to carry out word meaning disambiguation, linguistic technologies look up its lexicon to find all possible meanings. These lexicons are semantic networks.

Semantic networks are not plain dictionaries, but resources that have been optimized for programmatical use, where word forms are knots linked to each other by multiple links denoting semantic or lexical relations. For example, the knots "secret agent" and "spy" are linked by a semantic relation named "synonymy" (they have similar meaning), while "angel" and "devil" are linked by "antonymy" (they indicate opposite concepts).

Something impossible for all those systems considering words as simple character strings, not concepts.

Memory

When people read, they unconsciously carry out a series of tasks before reaching full text understanding. One of these activities is keeping in mind concepts encountered so far.

During document analysis linguistic technologies use a similar technique to detect the context, enable disambiguation and the consequent extraction of meaning with the best possible approximation.

Knowledge

The cultural education is a key element in order to understand what is being read. For example, when people with an average cultural background read a specialized text on Einstein's Relativity Theory, they can easily understand the words and the general pattern, but they cannot understand the meaning due to a lack in specific knowledge.

Using linguistic technologies, we can have something very similar to what we usually define as culture: a vast and balanced knowledge of the real world. This knowledge has been organized in descriptive rules which are applied during the analysis in a process that can be compared to the human "common sense."

A data processing knowledge base acts similar to any general-knowledge base: it can be increased, enriched and improved in the same way as our knowledge base does when we learn new things.

It's sufficient to decide to add new or specific information to the knowledge base through guided learning mechanisms.

Content Representation

The outcome of linguistic and semantic analysis is a conceptual map of the text contents, where:

- each concept is stored independently from the words used to represent it
- each agent is associated with the action carried out
- each object is connected with the related action

Even the main subject of the document (for example "sport" or "finance") as well as minor subjects ("tennis" or "stock exchange"), special data like dates and numbers and other meaningful information can be stored in this representation.

The fundamental feature of this conceptual map is a set of structured data. For this reason it can be used in any formal processing tasks such as indexing, classification, summarization and translation.

6.5 The Limitations of Other Technologies

The linguistic technology based on semantics is superior to other traditional technologies when it comes to building real knowledge management solutions instead of simply manipulating words. Let's have a look to the search engines, for instance. They are considered to be the most popular applications, undoubtably the most common ones.

Let's remember however, that the technologies for the automatic processing of unstructured information have been designed to be used in many other applications, besides the search, such as: content automatic classification and data extraction, natural language processing interfaces (the user poses questions in everyday language and the system returns the best possible answer), and automatic translation.

Linguistic Technology and the Full-text Systems

"Full-text indexing and retrieval" systems act on a "cleaned-up" version of the text, where have been deleted all the most frequently-used words without a proper meaning (such as articles and prepositions). In regards to this type of systems the document is only composed of character strings appearing several times.

This approach has its power in the indexing phase, which is simple and quick. However, its shortages make it less suitable in the advanced search for information because it causes overload and underload problems. Let's consider their main limitations.

Words = Character Strings

Let's consider this important declaration of a business person:

"This year we are closing two plants in Italy and opening another two in Poland."

Full-text systems index the words "closing" and "opening" literally, without recognizing they are forms of verbs "to close" and "to open". Nobody looking for information on plants shutdown will think about using words like "are closing"; therefore they can't find the document containing the sentence.

Considering the words as simple character strings, full-text systems skip a lot of information and do not analyze it. What's worse, they use a lot of CPU time to process irrelevant words — like conjugated verbs — and a lot of disk space to store them.

Some systems have a word-forms management that provides an automatic search of all the possible word variations, but many problems still remain even in this case.

For example, if you are searching for information about "knife edge", a system of this kind finds references to the word "knife" and to its plural form "knives", but still doesn't detect the location and so it misses the concept. Retrieving every document containing "knife", "knives", "edge" and "edges", it easily produces overload.

Linguistic technologies don't have these issues because they correctly recognize and automatically manage locations or collocations referring them to their exact meaning.

No Concept Management

Full-text systems cannot understand the meaning of the words they process, so they cannot collect any information regarding concepts.

For example, if you are looking for information about "driver" in the meaning of "person driving a bus", the only way to find pertaining documents is to specify both "driver" and "bus" as the search criteria.

This leads to underload, the system shows the documents containing both words first, placing documents that match only one word at the end of the list, even if they contain "driver" with the right meaning.

At the same time a situation of overload occurs because documents placed at the end of the list can refer to wrong meanings (driver = golf club, driver = computer software, etc.) and documents containing only the word "bus" are presented at the same level in the result list.

By representing the document content in form of concepts, instead, the system is able to avoid these kinds of problem.

Some systems try to avoid the above issue by searching synonyms of common words, so that users don't need to remember and write them.

This method lessens underload because it retrieves more possibly relevant documents; yet it considerably increases overload as it finds all the documents containing all words, independent of their meaning.

This example clearly describes the problem. The synonyms of "hold" are:

> *grip, grasp, clutch, clasp, clench, capture, taking, pinch,*
> *possess, bind, store, hug, footing, sustain, wait, power*

If the user is looking for "hold" meant as "capture", the use of synonyms increases the number of relevant documents, but at the same time it retrieves many more irrelevant documents that contain words having nothing to do with the searched meaning ("pinch", "grip" and so on).

The search for "hold", meant as "power", is even worse. The additional documents retrieved by means of "hold" synonyms are not relevant and the overload prevails. By storing concepts, linguistic technologies have no problems in managing these cases automatically and in a way that is transparent to the user.

Boolean Search Operators

Linguistic and semantic technologies are used in search engines that allow users to pose questions in everyday language, as if they were speaking to someone. For example, in case of a search inside a government document base, it is possible to pose questions such as:

What documents are required for a registration of a boat?

and obtain the right answer.

In full-text systems, users can try to obtain similar results by means of Boolean operators like "AND" and "OR":

(document OR documents) AND ("registration" OR
"to register")

Actually, this question is not equivalent to the first one, because the forms "document" and "documents" do not cover the whole concept (think of "certificate", "identity card", "passport", etc.) Both underload and overload situations may occur, because the references to "document" and "documents" can be recollected to the verb "to document."

Moreover, in the example the user should have entered the synonyms explicitly; an unlikely situation, since there would often be too many synonyms.

Writing search criteria with Boolean operators is not a common action because it is an unnatural process. Even experts often make several attempts before obtaining the desired result.

Regarding the full-text systems, some vendors boast about their ability in managing questions posed in natural language. Actually these systems simply use hidden Boolean operators.

They transform written questions into Boolean search criteria, removing the words with no proper meaning (for example "for", "the", "from"), which are ordinarily used ("and", "what"), obtaining the criteria like (only a partial list):

> documents AND required AND registration AND boat
> (documents OR required) AND (registration OR boat)
> documents OR required OR registration OR boat

Usually these workarounds provide low quality results, because only some pieces of relevant information are found (underload) together with many unrelated documents ("documents" and "registration" are very common terms in the document base) causing a considerable overload.

Sorting

A problem that needs to be solved when dealing with big document bases is the sorting of the search results. Ideally, results should appear ordered by relevancy, with the most relevant documents at the top of the list and the others following.

Full-text systems use a simple statistical approach in order to sort the results: the more the occurences of the interesting words, the more the importance given to the document. This method proves to be right in some cases, but often it is unable to provide the right sorting, forcing the user to waste time browsing all those superabundant results for the most appropriate document.

An example can better explain the issues of this approach. In a text dealing with "online purchase of regional food", the word "purchase" appears several times and the search for it, will place this document among the most important ones, while the main subject doesn't concern the purchase but the products described.

Linguistic-semantic technologies can solve this problem as they can express more clearly the search to be carried out (thanks to natural language comprehension). In addition, they use conceptual maps that do not refer directly to words but to interrelated synthesis information.

6.6 Linguistic Technology and Pattern-Matching Systems

Pattern-matching systems (or neural networks) is based on the same techniques of full-text products, but they use a more advanced statistical analysis on contents.

The ideas adopted in order to improve the system behavior are:

- fuzzy logic
- pattern matching
- word co-occurrence
- importance of the less frequent words.

Fuzzy logic

This name refers to the use of an intentionally "imprecise" logic, which has the aim of solving problems coming from processing words only as strings of characters.

The idea is to broaden the search criteria by adding words having the same root of that specified by the user, e.g., "bottle, bottles" when the user has specified only "bottle."

At first glance this solution seems ingenious but results often get worse. For example, in the case of the word "pressing" (adj. urgent), overload can occur, because the word "press" has the same root but different meanings (i.e., newspaper, ironing, machine use for printing).

Other problems show up with irregular terms, independent of their gender. For example, with nouns such as "knife" — pl.: "knives" — documents containing plural forms are not found, leading to an underload. The same happens with verbs such as "to go" — past: "went" — because the root is different.

Pattern matching

In this case the idea is to broaden the word reference unit to a group of words. Instead of considering each word as an independent object, this function considers the word group as unique objects that repeat themselves with a given frequency.

In this way, a collocation like "bird of prey" is managed as a unique form and not as broken words, improving the search capability.

However, the system stores the couple bird/prey as a unique object and therefore it also finds this group in a sentence (for example "Felines see the flying birds as prey") where the meaning is completely different.

In addition, the system keeps processing these objects as simple sequences of characters and not as concepts. Therefore it is unable to generalize their management: "birds of prey" is not detected as a form of "bird of prey", but as a different object.

Co-occurrence of terms

It has been observed that when many of the words in a group appear together in several documents, chances are high that those documents have the same subject.

Building on this principle, this method analyzes co-occurence of words inside texts statistically, making automatic classification possible. Yet, this method achieves its best in classification and doesn't appear to improve search.

Surely a statistical trend exists that links word occurrence to semantic similarity, but it cannot be generalized in an absolute and reliable way, so it proves to be only partially useful.

Let's imagine a system of this kind adopting a rule like:
Documents in which the words "score", "play" and "shot" appear together are about sport.

However, the words have different meanings according to the context and cannot be disambiguated without a context.

(a) *The Lakers make a great play and score with a shot from the 3 point line.*
(b) *He poured a shot of whiskey and started to write the score for the play.*

The words "play" and "shot" are nouns in both sentences, but have different meanings; "score" is a verb in (a) while in (b) it is a noun. For this reason, the system will fail classification of sentence (b).

In other words co-occurrence of terms can be useful in the search for statistically similar documents, but it doesn't enable to perform more accurate searches than traditional full-text systems.

Importance of less frequent words

Another functionality of these systems is a special management of words that are statistically less frequent. The theory behind this states that the most informative words of a text are the less common, and consequently must be given more relevance. This setting can make the document classification easier, because it exploits a statistically observed feature of texts, but it doesn't add value to the information search.

The information contained in a document is expressed also through the use of common words, because these can give meaning to sentences. A system that only considers less frequent terms as indicators of the content can never be used as an effective information management tool. The reason why writers use all those words (and not even one less) is because they all are necessary!

Since textual information is usually created to be read and understood by other people, an analytic approach, similar to the one used by humans, is the best option to obtain high quality results in the automatic processing of unstructured textual informations.

6.7 The Linguistic Platform of Expert System

COGITO® is the linguistic platform of Expert System, a set of technologies and proprietary resources, the result of hundreds of years of research and development.

COGITO® depicts the knowledge contained in texts written in the everyday language. In other words, it comprehends natural language, understands the meanings, translates in different languages, and shares the knowledge, etc.

COGITO® interprets the concepts contained in the texts, just as a human would do.

Thanks to the most advanced linguistic technologies, COGITO® recognizes all the aspects:

- structure
- vocabulary
- semantics

in the texts, through a deep analysis and disambiguation of all the elements, thanks to a rich and complex "real world" representation.

Unlike the other approaches of automatic processing, COGITO® does not just superficially manipulate a group of words which can result in "guessing" the meanings and concepts using keyword searches, lists of synonyms, and statistics, etc. Instead, COGITO® surpasses all others by facilitating understanding and conveying actual knowledge through its semantic network, which contains millions of pieces of information about terms, concepts, abbreviations, phraseologies, meanings, domains and connections among terms.

Features of COGITO®

COGITO® relies on a pragmatic architecture with features developed to support the specific needs of companies of any size. The platform is efficient and can quickly process large amounts of data and texts.

In addition, it is easily portable to different environments and operating systems.

Over time COGITO® has been integrated with the most popular products and platforms, proving day after day to be a reliable, robust and sound solution.

Elements required for processing natural language

COGITO® is composed of various modules dedicated to specific activities needed to disambiguate texts and process natural language which are essential for the automatic comprehension of questions formulated in the everyday language.

To understand automatically a text we need:

- a semantic network
- a parser to trace each text back to its basic elements
- linguistic engines to query the semantic network (to link the basic elements of the texts with the semantic network of the meanings)
- a system of disambiguation

COGITO® components

Expert System has developed the following modules:

(1) SENSIGRAFO
A set of semantic networks firmly connected in the form of a graph that contains the conceptual representation of the language.
The network contains:
— information about connections among objects
— specifications about the lexical domains of each word
— information about the frequency of use

1 A) GEOGRAPHIC SENSIGRAFO

A representation of all the significant geographic localities and the respective connections arranged in an intelligent way.

1 B) SENSIGRAFO FOR COMPANIES AND PRODUCTS

A system that understands subjects and concepts included in a document and distinguishes automatically common names from brands.

1 C) SENSIGRAFO OF NAMES AND FAMOUS CHARACTERS

More than 100000 characters, 3000 companies and 20000 localities.

(2) MORPHOLOGICAL, GRAMMATICAL AND SYNTACTICAL PARSER
The parser performs a complete morphological, grammatical and syntactical analysis of the sentence, managing more than 3500 rules very quickly. The parser uses an innovative and ad hoc methodology to query the semantic network, resulting in a significant improvement of the existing parsing methods.
(3) THESAURUS
The most complete existing thesaurus for the Italian and English language. When included in a search engine it permits the query of the database using the researched word connected to its synonyms.
(4) ADVANCED GRAMMAR CHECK
The rich underlying dictionaries make the Expert System advanced spell check the most accurate product on the market (it is the only one able to fully support inflected forms, verbs and enclitic particles, compound nouns, verb context, etc.)
(5) ITALIAN–ENGLISH / ENGLISH-ITALIAN DICTIONARY
The dictionary contains more than 150000 entries and 300000 phraseologies.

(6) ITALIAN VERTICAL DICTIONARIES
A single set of specialist lexica (IT, medical, economic, etc.) that analyzes specific texts of special sectors in an accurate and comprehensive way.

(7) AUTOMATIC CLASSIFICATOR
Using the sensigrafo and specific rules of elaboration, the classification technology automatically catalogues a document within a vast hierarchy of domains and sub-domains.

(8) MEANINGS DISAMBIGUATOR
For a human, the meaning is something obvious because of our capability to refer automatically to cultural elements that help us to understand the meaning of a word. Transferring these capabilities to a program or to a system of automatic interpretation and comprehension of the texts is among the main accomplishments of Expert System and the source of its competitive advantage. The disambiguator of meanings included in COGITO® thoroughly analyzes sentences or whole documents and distinguishes the right meaning for each element found, eliminating possible ambiguities.

(9) KNOWLEDGE EXTRACTION TECHNOLOGIES
The knowledge extraction technology of Expert System performs activities of normalization and rationalization of the documents, in addition to the retrieval of significant information from unstructured texts for applications of text mining, such as the automatic population of databases.

Linguistic analysis of the texts

CORPUS and TAGGING represent the "behind the scenes" work of automatic elaboration of the language, the basic pre-requisite of any processing of text.

Expert System selected a complete and balanced corpus able to effectively represent the various compositions of the texts available in the real world.

Then Expert System proceeded to the manual analysis of a sub-set (balanced as well) of the corpus and to the disambiguation of the exact meaning of each polysemic word.

Is the creation of all the rules of disambiguation needed to increasingly delegate to the software programs the task of analysing and solving the ambiguities found in the texts. These rules have become a collection of each single possible meaning of the words and concepts which are typical of the most common context of each meaning.

Expert System enhances continuously its reference corpus, which is extremely rich and composed of articles and texts taken from the most important encyclopedias, newspapers, magazines, novels, essays, leaflets, paying great attention to obtaining an optimal balance of knowledge.

The behavior of SEMANTIC NETWORK and PARSING

The components of COGITO® can work both alone and in collaborative mode.

The Sensigrafo is the SEMANTIC NETWORK that constitutes the knowledge base for all the analyses performed by the linguistic engines developed by Expert System.

Within the Sensigrafo, each syncon represents a knot in the semantic network and is linked to the others by semantic connections in a hierarchical and hereditary structure. Each knot is linked to the characteristics and meaning of the connected knots.

The PARSER is the engine that identifies the single elements that constitute a text, assigning them the exact logical and grammatical value.

Other existing systems lack the optimal integration between semantic network and parsing. COGITO®, however, has worked in this very direction achieving unique results.

Steps of the analysis:

- creation and maintenance of a semantic network to code knowledge
- reading of the text and extraction of the elements of the sentences (engine)
- comparison with the network of meanings
- retrieval of the concepts from the analyzed text
- returning of the meaning (not ambiguous) of the text.

The disambiguation of texts

Disambiguating, that means....

... receiving input texts and returning the same texts as output, where each term is marked with the concept it represents. From a computational point of view, it is a sequence of passages of analysis of the text and improvements in the interpretation of the concepts contained in it.

This is because a program must be provided with a univocal representation of the world, creating a system of reference to represent the equivalent of the human experience of the world: a generic experience, of course, not an individual one.

Disambiguating: this is the true problem in the automatic interpretation of texts. In order to distinguish between

The rust eats the tower
The knight eats the tower

a program must be able to "reason." It must be taught that a language contains many ambiguities a man can solve without problems.

But what a man knows because of education and experience, a software must deduce from the text automatically, relying on coded knowledge and advanced technologies.

The research and development of automatic systems for the linguistic disambiguation must solve a crucial problem: the administration of the number of existing combinations that can be generated when dealing with words and texts. These can be combined together in a very high number of ways, increasing exponentially.

The problem that disambiguation engines face more frequently is that of reducing drastically the number of possibilities, as beyond certain thresholds, the increase in precision even of only a few percentage points (when possible) is extremely costly (in terms of time and resources).

Another limit of the existing solutions is that they can not expand their capabilities from single samples to the universe of the language without incurring in often very high percentages of errors; such systems become quickly unstable and are not able to solve the problem of linguistic disambiguation for every language and/or knowledge domain.
The disambiguator of COGITO® is able to "reason" to distinguish between the various meanings of all the elements of a text, individuating the context in which they are included.

The difficulties of creating a system able to work on all the possible ambiguities of a language are proved by the poor results obtained by most of the systems currently on the market. The disambiguator developed by Expert System, is a unique product because it is able to reach the most ambitious result of linguistics: making a program that understands the sentences like a human would understand them, thanks to his experience and culture.

The disambiguator can work sentence by sentence or considering whole documents, according to the way it is configured. Distinguishing all the possible meanings of a text is just an additional, but extremely critical, step beyond the more common analyses: logical, grammatical, query of the Sensigrafo, domain analysis.

There are many examples of interpretations of words that we humans can take for granted but a program can not, including expressions meant in a figurative sense. Some examples of what the disambiguator can do?

Understanding univocally the following sentences:

Mary delivered a beautiful baby
Mary delivered a good project

The disambiguator suggests that in the first case we are dealing with to deliver as giving birth, while in the second case as handing in.

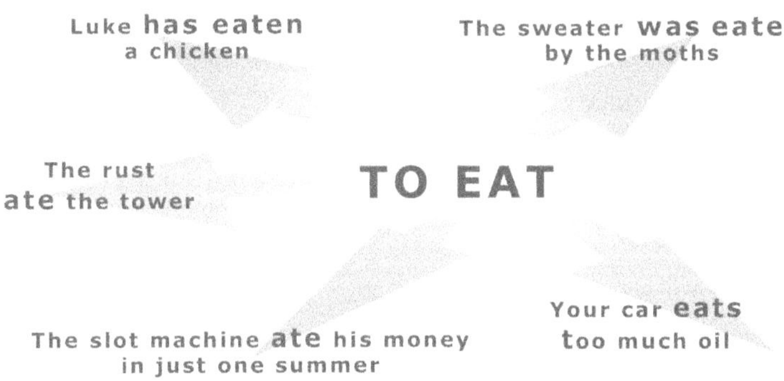

COGITO® surpasses other technologies

The system most used today for the comprehension and processing of the text written in natural language is the one known as backtracking, that is based on returning back to the previous steps of the analysis and modifying them, in case of events, during the analysis, that change the conditions and the validity of the previous assertions.

This method is generic and scalable but suffers more than any other approach from the problem of the exponential explosion of the tree of possible combinations. The challenge is that of maintaining the basic idea (that is, the possibility of interaction among different scenarios in order to choose the best one) while reducing progressively, the time needed for the analysis.

Precision of the disambiguation

- the actual systems, in free contexts, reach a level of disambiguation ranging from 45% to 55%, while some systems have reached a level that doesn't exceed 73–75%, if reducing the context;
- further reducing the number of reference meanings (with a gross interpretation of the meanings) they reach peaks around 80% (even if they are prototypes not suitable for real applications);
- Expert System thanks to its technology, in free contexts obtains a precision that exceeds 90%.

6.8 Semantics, Animals and Earthquakes

Semantic analysis allows the automatic comprehension of words, phrases, paragraphs and entire documents, and can contribute to the field of predicting catastrophic events, like earthquakes.

While there is no obvious connection between the field of semantics and earthquakes, semantic analysis, when combined with inferential algorithms that has been effective in predicting previous seismic events, can automatically analyze millions of pieces of data.

Semantic technology is able to:

- Understand all types of information contained in documents, emails and web pages
- Rapidly extract the key data and organize it in a database
- Discover the hidden relationships between the available data
- Expand on hypothetical scenarios and identify useful signals

There is a lot of unstructured information (such as personal accounts not organized in a database) available on earthquakes that is difficult to measure with classic scientific instruments. Through semantic analysis, important information can be extracted and organized from this unstructured data to provide more useful clues to better study seismic activity.

Considering all the knowledge gained through observing the strange behaviors of animals (chickens, cows, horses, dogs, birds, etc.) around earthquakes, is it only urban legend? While this phenomenon may not be supported by science, the fact remains that centuries of amazing anecdotal information exists to support the theory.

To this end, we suggest an interactive, open channel of communication where people can contribute their personal stories and experiences in observing strange animal behaviors around earthquakes. Available through a website, or via email or SMS, this system would allow individuals to record that their horse became nervous and escaped from its pasture, or that their hens did not lay eggs in the hours before an earthquake. This information could then be connected to existing data on seismic activity — historical information, recorded gas emissions, atmospheric events, etc., as possible predictive signals of earthquakes.

It is important at this point to use semantic technology, which is able to automatically and rapidly understand what is written in documents or texts. Using the following text, for example:

"*In my home, located in Spoleto, on the morning of May* 7, 2006, *my dog started to bark and growl, for no apparent reason, and became agitated, running up and down through the yard....*"

The information can be transformed into a database record, extracting the key details of the entry:

- Date: 5/7/2006
- Hour: 6:28
- Latitude: 42.7
- Longitude: 12.73
- Animal: dog

All of the witness accounts could be recorded in the same manner, in a database which then applies inferential algorithms in such a way that rules can be created that, with a good approximation, in terms of time, place and intensity, could provide advanced warning of impending disasters. The system would function as follows:

- People submit reports via a website, email or sms.
- Using semantic analysis, the system interprets and filters the information.
- A database is created that is cross referenced with other scientific data.
- The data is referenced through inferential algorithms to correlate the findings.
- Based on the analysis, the signals can be evaluated, and warnings can be issued, if necessary.

7

Satellite Sensors

Pietro Finocchio*, Antonio Lanzillotti†, Attilio Vagliani†, Daniele Brotto†, and Luca Pietranera†

**AFCEA-Rome Chapter*
†*MoD — Italy*

"It is funny to see that almost all men of great value have simple manners; and that almost always simple manners are taken as indicative of little value."

Giacomo Leopardi, Pensieri.

7.1 Aim

The Chapter describes the satellite sensors that can be useful in a space-based assistance to seismic prone territories. In particular the Synthetic Aperture Radar is dealt with.

7.2 Remote Sensing Systems

7.2.1 Introduction

Remote Sensing is a set of techniques which can capture information about an object without coming into physical contact with it.

It is obtained by measuring changes in an appropriate field (scalar or vector) determined by the presence of the object and/or changes in its state. In the follow on we will refer only to technologies based on measurements of the electromagnetic field. In this case, the electromagnetic radiation generated by an appropriate source (or the object itself) interacts with the object (the

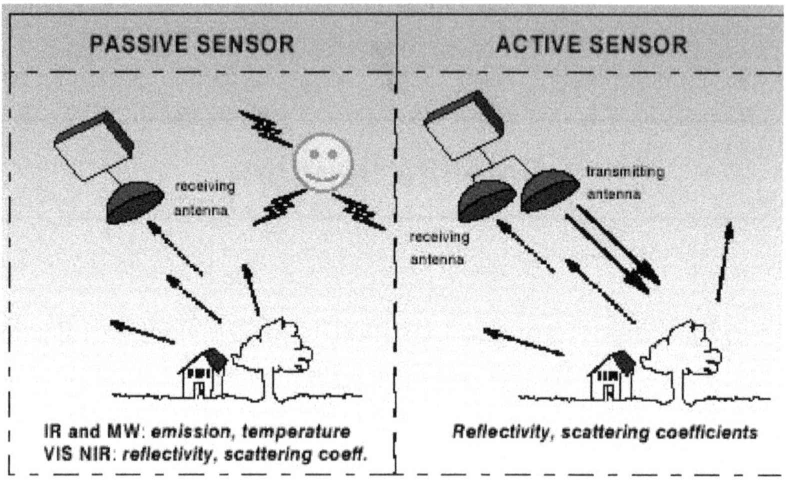

Fig. 7.1 Passive and active sensors.

phenomena of reflection, diffusion, absorption, etc…), and therefore is modified. This radiation propagates and is measured and recorded by a remote sensor.

Remote Sensing of the Earth's surface from space, for example, is obtained through the interaction of electromagnetic radiation, at different wavelengths, with the surface it strikes.

The remote sensing systems can be classified into "active" if the instrument emits electromagnetic energy receiving the backscattered fraction by the object under observation, "passive" if the instrument is only receiving and measuring electromagnetic energy emitted spontaneously from the means observed.

Analyzing and interpreting these measures of the electromagnetic field, it is possible to trace the properties of the object of interest. Other types of fields are used for specific geophysical applications, such as gravity, Earth's magnetic field and the air pressure (acoustic wave).

Remote sensing using the microwave band of the electromagnetic spectrum (1–60 GHz) can be regarded as a fundamental tool for the management of land resources, complementary to optical techniques in the visible and infrared electromagnetic field, especially for specific information that they are able to provide and for their ability to operate day and night in virtually any weather conditions (and therefore also in the presence of cloud cover).

7.2.2 Radar

The word Radar is the contraction of "Radio Detection and Ranging." A radar system is an active electronic system operating in the microwave field, able to detect an object, indicating its distance (range) from the radar system.

The term radar is also commonly used for all types of active microwave systems although they may not, in some cases, provide information on the distance of the target. The radar is essentially composed of an antenna, a diplexer, a transmitter, a receiver, a signal processing system, a data presentation system.

The antenna operates normally both in transmission and in reception (monostatic radar). The transmitter generates the electromagnetic field of Radio Frequency (RF) with a power level suitable for sending to the antenna. This RF signal is typically appropriately modulated (in amplitude or frequency) in relation to the particular radar technique used to discriminate between the various elements of the surface under observation. The receiver converts the signal back to the antenna at lower frequencies and reveals it. Very often, in addition to measuring the level, it is also able to measure the phase with respect to that of the transmitted signal. In such a case we speak of coherent radars to distinguish them from non-coherent radars, which only carry out power measurements. Many of the radars used in remote sensing are coherent.

7.2.3 The Radars for Remote Sensing Applications

The radar sensors most commonly used in remote sensing are the scatterometer, the radar altimeter and the SAR (Synthetic Aperture Radar).

Amongst these, the scatterometer is the one specifically dedicated to measure the radar cross section with high accuracy. The prevalent applications are in the experimental measurement of scattering properties of the ground and the study of patterns of interactions between electromagnetic radiation and natural means. For these applications scatterometers are normally used which are either land based or installed on aircraft. A special type of scatterometer, equipped with multiple antennas, is also used on satellites to measure wind speed and direction on the sea's surface. Although it provides bi-dimensional measures (wind field), they are not considered real images. The radar altimeter is used on satellite and carries out measures almost over a small area aligned along the satellite ground track. They provide, if properly interpreted and processed, oceanographic information, such as wave height, wind speed over the sea, the topography of the sea surface (geoids, currents, etc.).

The SAR is, instead, expressly designed with the goal of producing microwave images with high spatial resolution.

7.3 Synthetic Aperture Radar (SAR)

7.3.1 A Historical View

The Synthetic Aperture Radar (SAR) was developed, starting in 1951, following observations made by Carl Wiley of the Goodyear Aircraft Corporation. The first radar images made by an aircraft were produced by a group of people belonging to the University of Illinois on June 8, 1953. Subsequent research by the Goodyear Corporation and the University of Michigan significantly improved the resolution of radar images produced in flight. This new technology, initially appreciated for its military applications, was considered restricted until 1964.

The first non military system, able to produce radar images, was built in 1968 by Westinghouse Electric Corporation and Raytheon.

The American province of Darien was fully shot This had never been achieved before, because of the constant cloud cover over the area.

The enormous technological development occurring in the following years led to the creation of an L-band SAR system, hosted in 1978 on the SEASAT satellite. This was the result of a multidisciplinary effort led by the NASA JPL (Jet Propulsion Laboratory), with the active participation of NOAA (National Oceanic and Atmospheric Administration).

After this first experimental observation of the Earth, starting in the early nineties, almost all the Space Agencies have included, in their programs of Earth remote sensing, the launch of platforms with SAR sensors on board. The main missions conducted in the past were programs ESA ERS1/2 and ENVISAT, the programs SIR-C/X-SAR and SRTM, who have seen the use of SAR on board the shuttle bus, the Canadian RADARSAT 1 and 2.

The SAR has been used successfully in the field of planetary exploration with NASA's Magellan mission, allowing us, for the first time, to get a radar map of global 3D surface of the planet Venus, whose dense atmosphere is impenetrable to optical radiation. The Magellan probe (see Table 7.1 below) launched in 1989 orbited Venus from 1990 to 1994. It was the first of three

Fig. 7.2 Magellan probe.

probes launched to reach other planets from the Space Shuttle (the other were the Ulysses and Galileo) and was also the first spacecraft to use aero-braking techniques to lower its orbit. These techniques are used on current missions around Mars, allowing large quantities of fuel to be saved.

Since 2002, the interest in the applications of SAR on board of satellites has allowed, in the European context, the development of the Germany's SAR-Lupe (exclusively for military uses) and TerraSAR-X (for civil uses) as well as Italy's dual use program (i.e., inherently conceived and developed for dual use, both civil and military) COSMO-Sky Med, which to day and in the near future, can be considered the flagship for Italian technological excellence in this sector at global level.

7.3.2 Theoretical Principles

Let us consider a traditional monostatic radar (i.e., transmitter and receiver are both present in the same position). The transmitter sends a series of pulsed signals lasting a time τ, centered at an instantaneous frequency f and having

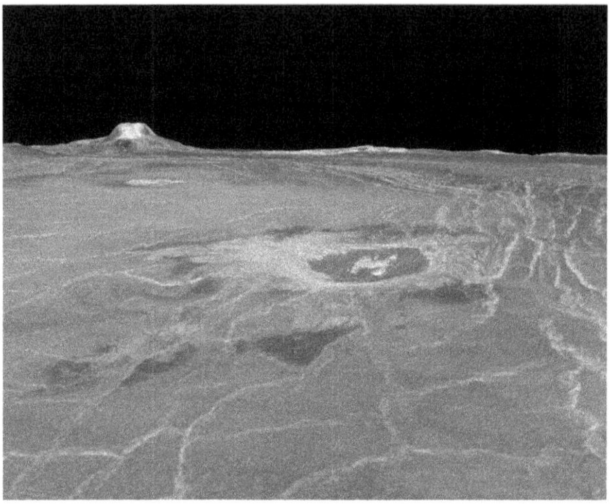

Fig. 7.3 3D view of Eistla Region on Venus derived from Magellan radar data.

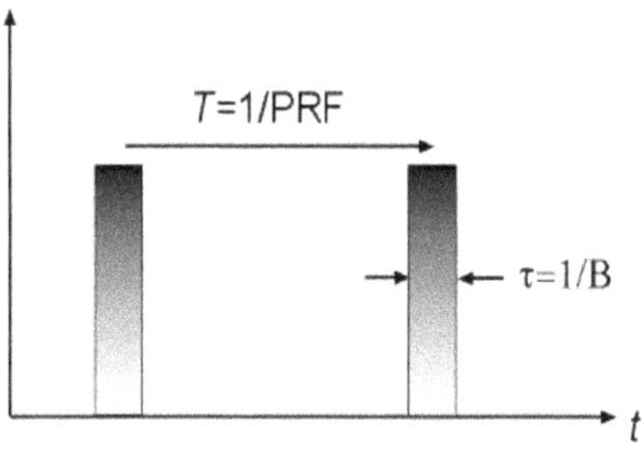

Fig. 7.4 Monostatic radar.

its bandwidth B. They are also equally spaced in time according to the PRF (Pulse Repetition Frequency) of the Transmitter (Figure 7.4):

The term spatial resolution is defined as the minimum distance between two points that the sensor sees as separate. We can distinguish between distance resolution (or range) and azimuth resolution. The radiometric resolution is the minimum variation of the signal perceived by the radar. This, as we shall see later, is an important parameter in defining the quality of the scene reconstructed from the backscattered signals received by the radar.

The range resolution r_r (which corresponds to the primary radar cell size) along the direction of the radar beam (slant range) depends on the duration τ

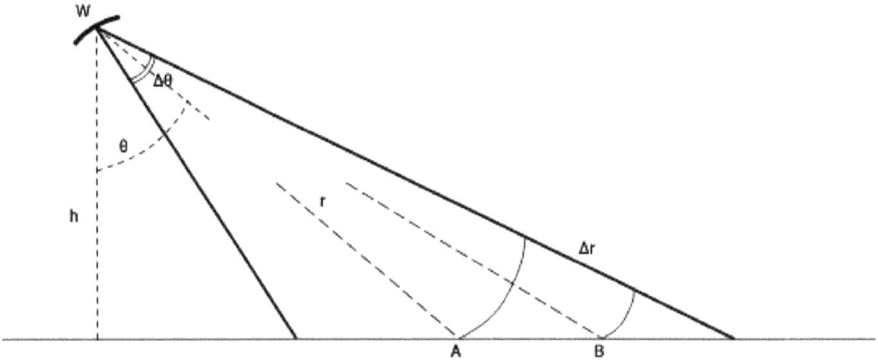

Fig. 7.5

of the pulse through the relation:

$$r_r = \frac{c\tau}{2} \approx \frac{c}{2B} \tag{7.1}$$

with $c = 3 \cdot 10^8$ m/s is the light speed in the vacuum.

The relationship referred above can be obtained using the following considerations.

Let us consider A and B two points (objects, small areas etc...) in the illuminated region, whose distances from the radar antenna are respectively r_A and r_B. If the signal has a pulse duration τ, the objects are resolved if the backscattered echoes do not overlap.

If $r_A < r_B$ the backscattered signals from points A and B are respectively received after a time $\frac{2r_A}{c} e \frac{2r_B}{c}$. The idea of separation of radar echoes can then be expressed as:

$$\frac{2r_A}{c} + \tau \leq \frac{2r_B}{c} \tag{7.2}$$

Namely:

$$\Delta r = r_r = |r_B - r_A| = \frac{c\tau}{2} \tag{7.3}$$

But, to create a map of the Earth's surface we must consider the range resolution projected on the surface, i.e., its projection on the ground (ground range), which is equal to:

$$r_{gr} = \frac{c\tau}{2 \cdot \sin\vartheta} \tag{7.4}$$

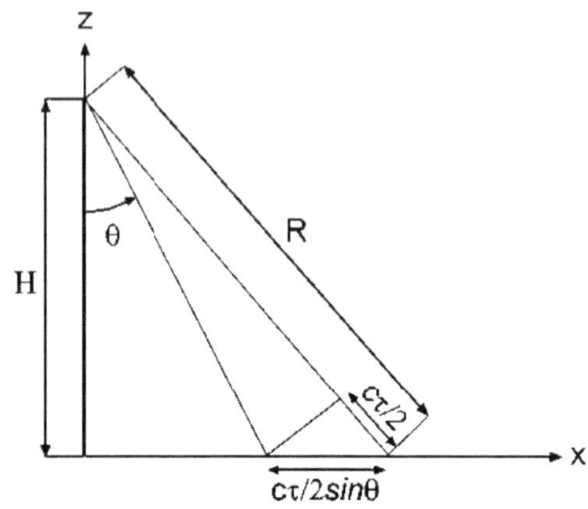

Fig. 7.6

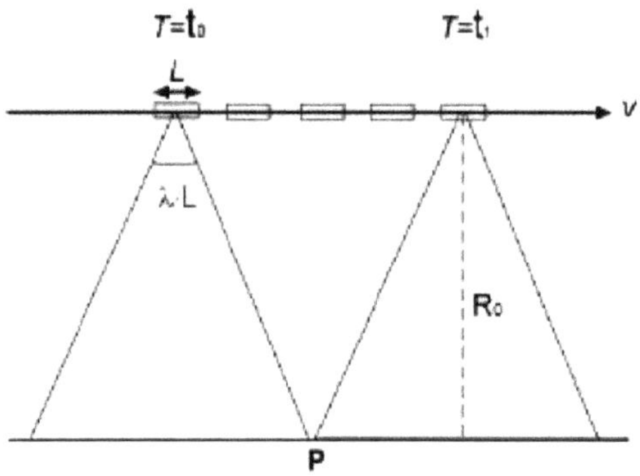

Fig. 7.7

where θ is the angle of incidence of the radar beam in the distance, as reported in the Table 3.3 below:

In the azimuthal direction, instead, the resolution achievable is equal to:

$$r_{az} = \beta R_0 = \frac{\lambda}{L} R_0 \tag{7.5}$$

where L *is* the length of the radar antenna in azimuth, β is the opening of the radar beam andR_0 the distance between the satellite and earth's surface (Figure 7.7):

Since the pulse duration τ is of the order of a μ sec, the resolution in range is of a few km; in addition, in order to avoid that the resolution along the Earth's surface is not much worse than the range resolution, it is necessary

Fig. 7.8

that the value ϑ is as close as possible to 90°, which means that the sensor has to illuminate the Earth's surface with a high grazing angle.

Therefore, as we have seen, improving the resolution in range, it is necessary, firstly that: the sensor illuminates the Earth's surface with a high grazing angle and secondly, to reduce the value of τ using a pulse of shorter duration, i.e., in the order of 10^{-7}, 10^{-8} sec., which corresponds to a very wide bandwidth of the transmitted signal. However, it must be remembered that the detection capability of the sensor is directly proportional to the energy E of the transmitted signal, which is equal to $E = P\tau$. Since the maximum power P is limited by the hardware of the radar, in order to increase *the* value of the energy E we should increase the pulse duration τ.To meet these two apparently conflicting requirements (increased resolution and high energy), the trick is to use special waveforms with pulse compression, called chirp (Table 3.5):

whose mathematical expression is the following:

$$x(t) = \cos\left(2\pi f_0 t + \frac{\alpha t^2}{2}\right) \cdot rect\left[\frac{t}{\tau}\right] \tag{7.6}$$

As far as the azimuth resolution is concerned, for example, for radar installed on a satellite in low Earth orbit, at a distance of 1500 kilometers from the surface and an antenna beam width of 0.1°, the resolution is about 2500 m, a too high value not suitable for applications.

To obtain azimuth resolution of a few meters from satellite, it would be necessary to have an extremely narrow beam, for example a width of about 0.0002° to achieve a resolution of 5 m, but these would require antennas wide 70 km in L band and 10 km in X band, values. This is clearly out of the question from the technological point of view. Therefore, a method to obtain high azimuthal resolution with restrained antenna aperture is to apply the technique of synthetic antenna, which radar SAR is based on.

7.3.3 Working Principles

The basic principle of SAR is as follows: if we consider a certain area to explore and we store a series of radar echoes obtained during the area's flight over, we may be able to get an azimuthal resolution (or cross-range, i.e., in the motion's direction) equal to the one of an antenna whose opening is equal to the distance covered by the aircraft or the satellite when detecting the echoes. The echoes backscattered from the area of interest while the platform is flying over it are produced by pulses sent from different antenna positions. For each and every position of the antenna, the radar transmits a pulse and stores the received signals into a computer.

The opening is thus obtained by sequential synthesis (SAR: Synthetic Aperture Radar) rather than instantaneously as in real aperture systems (RAR: Real Aperture Radar.) In evaluating the SAR impulse response (and therefore the resolution) of the SAR the following simplifying assumptions are considered:

- *start-stop operation*: the antenna remains stationary during the return time of the pulse, then goes immediately to the location of the next pulse transmission. This hypothesis stems from the fact that the speed of the platform is much less than the speed of light, so it's absolutely right to assume that the antenna's displacement during the time between transmission and reception of the pulse is negligible. It's just for this reason that in the case of SAR we can use the same equations of monostatic radar;
- *rectilinear geometry* (also called hypothesis of a "flat- earth"): the antenna moves along a straight line and the remote sensed surface is flat and parallel to the trajectory of the antenna;
- *temporal coherence of the scene*: the characteristics of the remote sensed surface's backscatter do not change during the period of observation;
- *ideal propagation*: there are no phenomena of attenuation, refraction, etc.

Consider then a satellite in orbit characterized by a velocity field $v(t)$, with a real antenna long L along the azimuthal direction. Assume further that the antenna radiates isotropically. At the time t the radar beam starts to illuminate the point P (Figure 7.9).

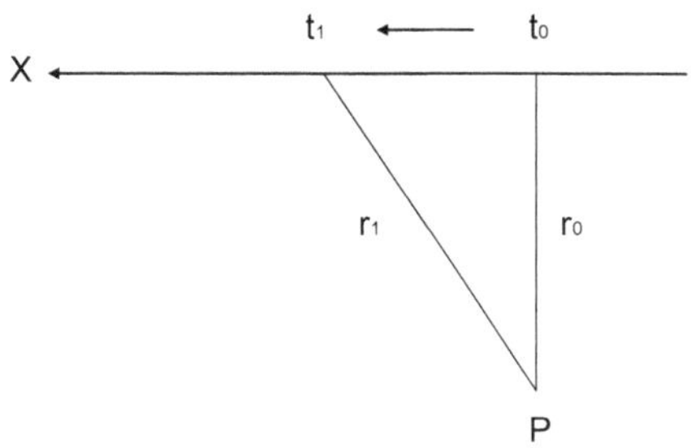

Fig. 7.9

In its orbit the transmitter sends a sequence of pulses with a certain PRF (Pulse Repetition Frequency) that hit P during the time interval $t_1 - t_0$. The real antenna then occupies a series of N equally spaced positions, each of which sends a pulse to the target. The antenna L then defines a synthetic array of length L_s along the azimuth direction or, equivalently, a single synthetic antenna long L In the time interval $T = t_1 - t_0$ the satellite covers the distance:

$$L_s = v(t) \cdot T = \frac{\lambda}{L} \cdot r_0 \tag{7.7}$$

defined as the "SAR azimuth antenna's footprint".

Since

$$r_{az} = \frac{\lambda}{2L_s} r_0 \tag{7.8}$$

at the same time we obtain:

$$r_{az} = \frac{L}{2} \tag{7.9}$$

i.e., there is an apparently surprising result, namely that the resolution in azimuth is independent both from the distance (flight altitude) and the wavelength. It depends only on the physical size of the antenna and is better, the smaller the antenna, contrary to what occurs in traditional radar. Nevertheless it is not possible to reach arbitrarily small values of resolution, because the antenna size reduction also reduces the gain and worsens the ratio peak/lobes, so the length L is typically on the order of meters. Another problem encountered is the migration in range of the resolution cell. As a matter of fact, in the time interval T during which the satellite runs along its orbit, the position of the illuminated point varies, due to the Earth's rotation. In the moment the illuminated point comes out of the resolution cell in range/azimuth, the SAR illuminates another point.

7.3.4 Acquisition Techniques

The acquisition modes of a SAR sensor are typically the following: Scan-SAR, StripMap and Spotlight.

In the ScanSAR mode, the SAR has a very wide swath with a particular type of periodic movement of the antenna pointer that also illuminates different smaller sub swaths. This unique feature makes the scan-mode SAR very helpful.

In the StripMap mode, the most common mode in the SAR sensors, the antenna is pointed along a direction fixed with respect to the flight path. Due to the displacement of the platform, the antenna footprint covers a strip that is limited in range and unlimited in azimuth.

Finally, in the Spotlight mode, the sensor drives the antenna's beam in order to continually illuminate the portion of ground to be detected. This mode offers an azimuth resolution better than the one Strip Map mode with the same antenna.

In Figure 7.10 the three SAR modes are shown.

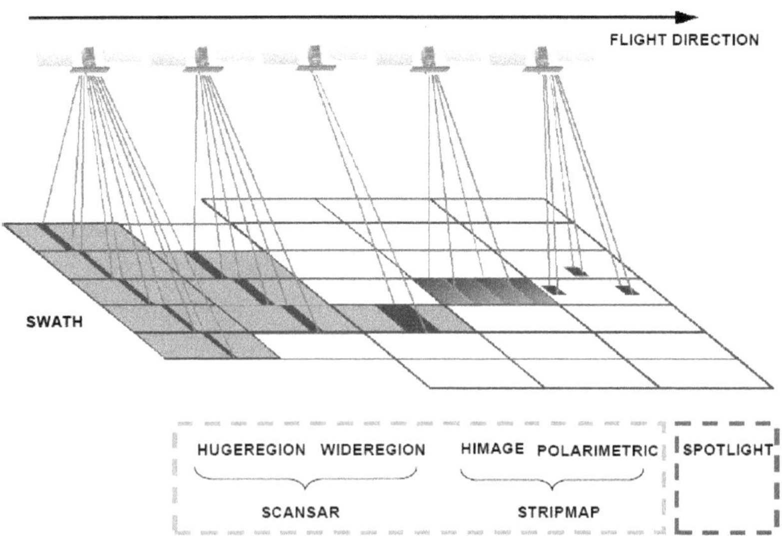

Fig. 7.10

7.3.5 Characteristics and Structure of SAR Imagery

The advantages of SAR compared to the usual optical systems are linked to the ability to operate at night and in the presence of clouds (there are regions of the earth where no optical images from satellite are available because of the perpetual cloud cover), yet, the SAR can provide coherent images that is to say that they contain both the information of intensity (related to the reflectivity of the objects) and the phase information (related to the distance between target and radar). A visible example of the difference between SAR and optical image which is illustrated by the following Figure 7.11:

A fundamental difference between the two types of images, which seems to be universally true, is linked to the fact that they are on different bands of the electromagnetic spectrum. Indeed, while an optical image can be interpreted in the light of our own sense of sight, the radar image has properties that, even when compared to the visual sense, they seem even counter-intuitive. In the interpretation of SAR images, it is extremely important to understand the effects such a type of shooting system introduces, which make it substantially different from the optical systems. The SAR images are in digital form and are constituted by a set of individual pixels (picture elements). Each pixel refers to the contribution by the individual reflecting point on the surface

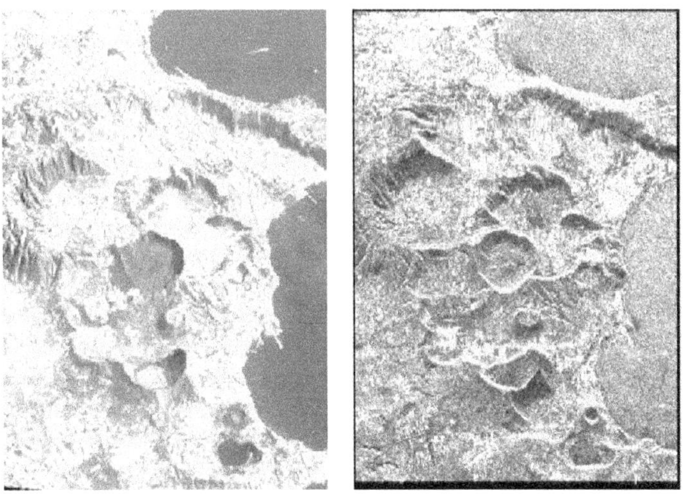

Fig. 7.11 Comparison of an optical image (SPOT satellite) and SAR (ERS-1 satellite) of Campi Flegrei area (Naples).

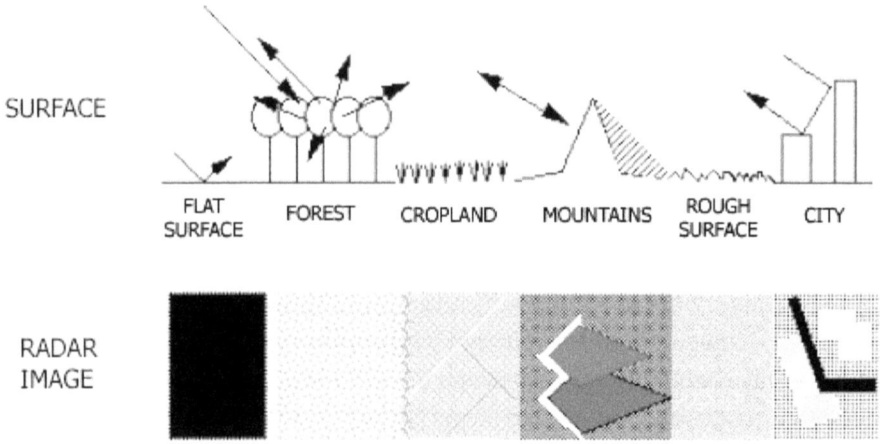

Fig. 7.12

illuminated by the beam. The amount of backscattering determines the level of gray (tones) of the object: for example, water constitutes a relatively dark tone as it produces little backscatter toward the radar while the vegetation is characterized by relatively brilliant gray tones. In general, bright colors mean that a large amount of radar energy is reflected back to the radar itself, while the darker tones imply that little energy was reflected. Echoes are a function of many conditions, such as wavelength, scattering object size, moisture content of the illuminated area, polarization of the transmitted energy, grazing angles (Figure 7.12):

In the follow on, this issue will be investigated through an analysis of some images supplied by the satellites of the COSMO-Sky Med constellation.

7.3.6 Speckle and Radiometric Resolution

The advantage of having a resolution independent from the flight altitude is paid for through the noise. The images acquired by the SAR are affected by the phenomenon of speckle (the same effect can be seen pointing a laser on a wall not perfectly smooth), which is defined as the image's intensity random variation around the average value $\bar{x}$ of the coefficient of backscatter. This variation is due to the random combination of several radar returns of individual scattering objects contained in the resolution cell. So homogeneous areas are presented as a sequence of small areas because the backscatter coefficient

Fig. 7.13

is a random variable x whose dispersion around the average value can be succinctly characterized through the coefficient of variation C_v equal to:

$$C_v = \frac{\sigma_x}{\bar{x}} \tag{7.10}$$

defined as the ratio between standard deviation and mean value. The coefficient of variation C_v thus provides a measure of the speckle, the greater is C_v, and the greater is the granularity of the image.

To reduce this phenomenon, specific techniques of signal processing are used. Unfortunately the use of such techniques is at the expense of geometric resolution. An example of reduction of speckle is shown in Figure 7.13 (ASAR sensor on board of ESA ENVISAT satellite):

The radiometric resolution Γ is defined by the following expression:

$$\Gamma = 10 \cdot \log_{10}(1 + C_v) \tag{7.11}$$

and constitutes an alternative way to assess quantitatively the speckle.

7.3.7 SAR Geometry — Distortions

Due to the acquisition geometry, a SAR image inherently contains a series of distortions. Therefore, a full understanding of the geometric properties of a SARs image can not ignore the knowledge of these distortions, both in range and in azimuth. From a general point of view, the distortions in range are much more evident than in azimuth and are mainly induced by the topography of the observed scene. The distortions in azimuth, instead, are much smaller, but their processing is very complex. In the follow on the main distortions in range will be illustrated.

Radar sends a pulse and measures the return time. Therefore, two radar returns from two scattering points placed at different heights reach the receiver at different times, and are seen as two points at different distances from the sensor. Thus the information related to the altitude is not known but is seen as different sensor-target distance. In calculating the resolution in ground range in paragraph 7.3.1 it was made the assumption that the angle α of the local ground inclination was zero. In a real case the formula for the resolution in ground range is:

$$r_{gr} = \frac{c\tau}{2\sin(\vartheta - \alpha)} \tag{7.12}$$

Concerning this, three types of distortion effects are commonly encountered when the observed scene belongs to mountainous regions, i.e., the effects of foreshortening (perspective view), layover (overlap), shadowing. In the first case points a' and b' in the radar image look closer (Figure 7.14):

In the second type of effect related to steep slopes, the SAR interprets the return of the peak as a return from a place at zero altitude, but at a closer distance in relation to the bottom of the protuberance of the ground itself. As shown in the Figure 7.15, the point b is closer to the radar then the point a, but the images a′ and b′ are swapped in the SAR image.

In the third type of effect, a part of the ground is shadowed and that section has no corresponding SAR image (Figure 7.16).

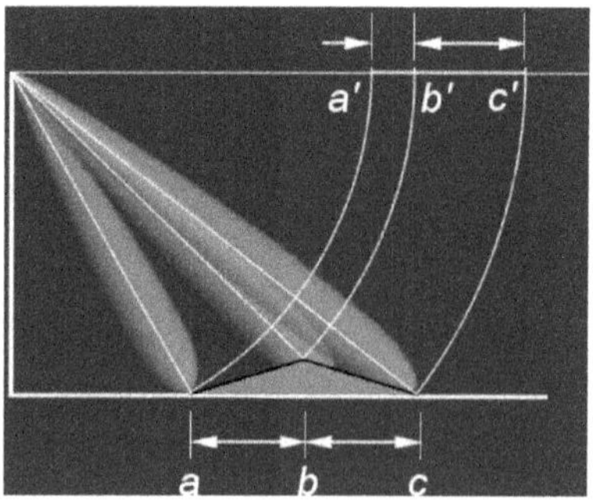

Fig. 7.14

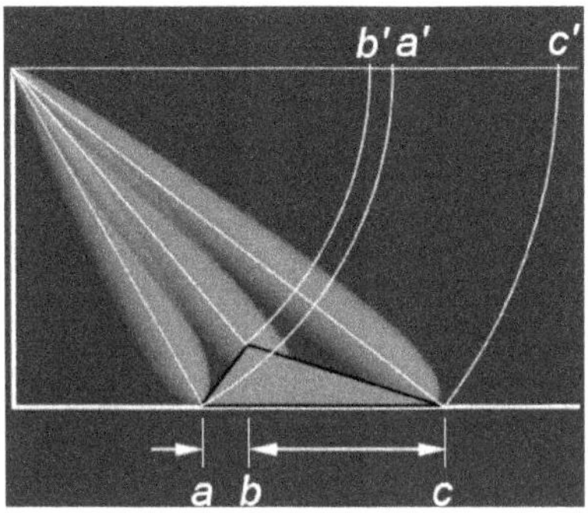

Fig. 7.15

Fig. 7.16

The comparison between the satellite ESA ERS1 SAR image of Mount Subasio area(Umbria) and a DEM (Digital Elevation Module), derived from cartographic data, shows the distortions present in the second image (Figure 7.17).

The foreshortening effects can be corrected by the geometric and radiometric calibration, assuming the availability of a DEM (Digital Elevation Model), while layover and shadowing can be calculated but not exactly corrected.

Another example of the strong geometric deformations that can be found in a SAR image is shown in this image of the skyscrapers in downtown Singapore, taken on 15th May 2008 with the COSMO-Sky Med satellite 1 in the Spotlight mode (Figure 7.18).

7.3.8 Geometric and Radiometric Calibration

With the term geometric calibration (or geo-codification or geo-referencing or ortho-rectification) is intended the process of converting a SAR image in an

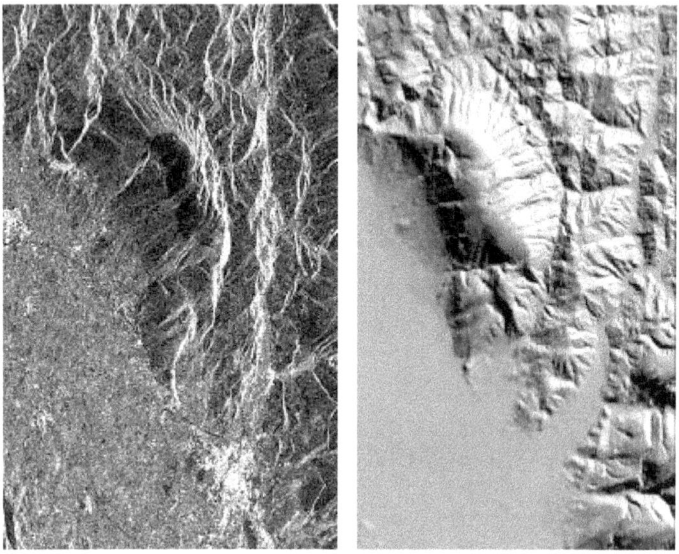

Fig. 7.17 On the left and right image DEM and ERS-1. Note the strong geometric distortion in the mountainous area.

Fig. 7.18

appropriate local coordinate system (such as a cartographic reference system) on the earth surface, useful for applications. The geometric calibration may be conducted with or without the help of a DEM.

With the term radiometric calibration is intended the process through which the sensor is calibrated with respect to a "standard" sensor. Calibration is necessary because only through that is it possible to compare SAR images obtained by different sensors or even images obtained with the same sensor, but in different shooting modes or processed with different processing algorithms.

For this purpose, corner reflectors are used, i.e., highly reflecting metal surfaces built and assembled to reflect almost entirely the beam (Figures 7.19 and 7.20).

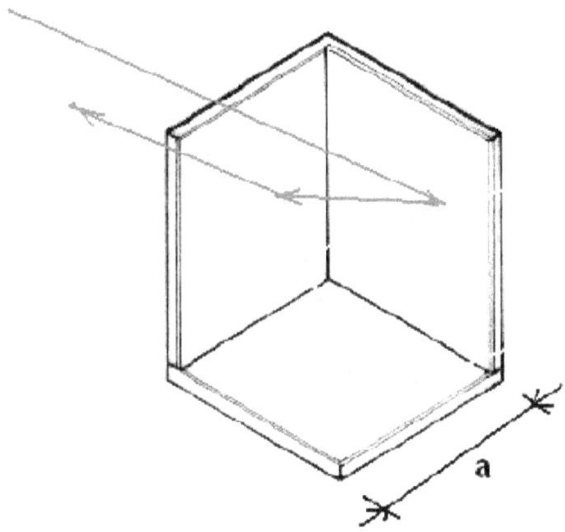

Fig. 7.19 Schematic drawing of a corner reflector.

Fig. 7.20 Corner reflector.

7.4 SAR Interferometry

7.4.1 IF SAR or IN SAR

Thanks to the applications of interferometry techniques in SAR systems (previously developed in radio astronomy), a method of treating the SAR data to calculate the phase differences, pixel by pixel, between two images of the same scene captured in similar geometric conditions has been derived. The radar signal is transmitted with a certain phase, recorded by a local oscillator on board. The signal, backscattered from the surface, reaches the receiver altered in amplitude in phase. The phase contains information concerning the return path. If two satellites S_1 and S_2 pass over the same area at different times (dual pass mode) (Figure 7.21).

The backscattered signals have the following complex envelope:

$$\begin{cases} S_1 = A_1 \cdot \varphi_{r_1} \cdot e^{-j\frac{4\pi}{\lambda}r_1} \\ S_2 = A_2 \cdot \varphi_{r_2} \cdot e^{-j\frac{4\pi}{\lambda}r_2} \end{cases} \tag{7.13}$$

where A_1 and A_2 are the amplitudes, and there are two phase terms, one proportional to the distance travelled by the radar signal (propagation phase)

$$\frac{4\pi}{\lambda}r \tag{7.14}$$

and the other related to the properties of the surface backscatter (φ_r: backscatter phase), subject to statistical fluctuations that result in an image, as known, affected by the speckle.

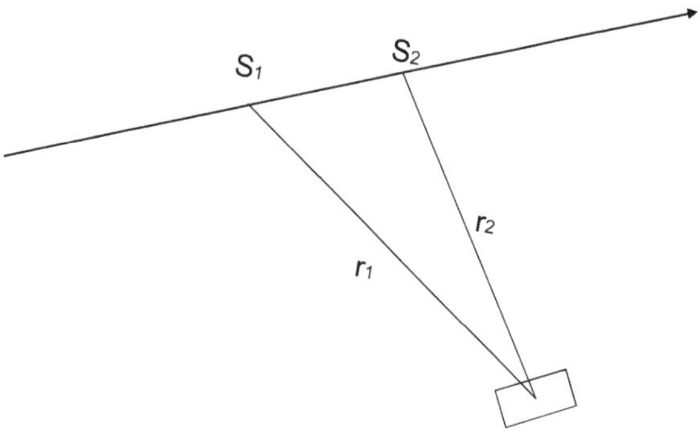

Fig. 7.21 Dual pass mode geometry.

The following quantity is defined as interferometric phase or interferogram:

$$S_1 \cdot S_2^* = A_1 \cdot A_2 \cdot |\varphi_{r_1}| \cdot |\varphi_{r_2}| \cdot e^{j(\angle\varphi_{r_1} - \angle\varphi_{r_2})} \cdot e^{-j\frac{4\pi}{\lambda}(r_1 - r_2)} \tag{7.15}$$

i.e., the map pixel by pixel of phase differences between signals S_1 and S_2.

The changes in properties of the surface backscatter observed between the two SAR transits are evaluated by defining the cross-correlation coefficient ψ:

$$\psi = \frac{E S_1 \cdot S_2^*}{\sqrt{E S_1^2 \cdot E S_2^2}} \tag{7.16}$$

The term $|\psi|$, called coherence, is used in the InSAR technique to indicate how much are varied the two SAR scenes which the interferometric phase is calculated of.

The interferometric phase is composed by several contributions: the first, referred to as flat ground phase φ_f, is the one due to the different angle of view of two satellites S_1 and S_2 (Figure 7.22)
From the above figure we can derive that:

$$r_2 = \sqrt{r_1^2 + B^2 - 2r_1 B \cdot \cos\left(\frac{\pi}{2} - \gamma\right)} \simeq r_1 - B \sin\gamma \tag{7.17}$$

where B is the baseline and then:

$$e^{-j\frac{4\pi}{\lambda}(r_1 - r_2)} = e^{-j\frac{4\pi}{\lambda} B \sin\gamma} = e^{-j\varphi_f} \tag{7.18}$$

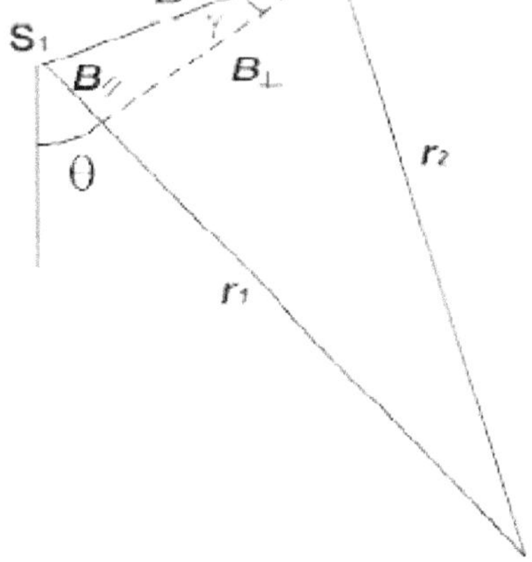

Fig. 7.22 Angle of view in the dual mode geometry.

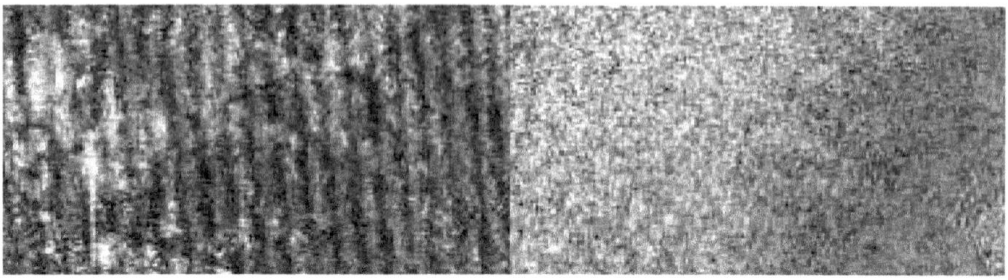

Fig. 7.23 Left orbital interferogram with fringes (flat earth), right after their removal.

This term means that on a flat surface, the interferometric phase is not constant but varies with the pixel distance from the sensor (range), giving rise to the so-called orbital fringes. These fringes depend on the relative positions of the two satellites.

So, knowing the precise orbits, they can be subtracted from the interferogram.

The removal of the orbital fringes will be as the better as more accurate is the information on the position occupied by the two satellites.

The formula shows that the contribution of flat land is directly proportional to the perpendicular component of the baseline, and then the frequency of the fringes increase as a function of orthogonal baseline (Figure 7.23).

In addition to the period of flat land, the interferometric phase is characterized by a series of contributions, according to the following formula:

$$\varphi_{\text{int}} = \varphi_f + \varphi_{\text{topo}} + \varphi_{\text{displ}} + \varphi_{\text{atm}} + \varphi_{\text{err}} \tag{7.19}$$

where:

- φ_{topo}: is the phase component that contains the topographic information, i.e., the relationship between phase and altitude;
- φ_{displ}: is the phase component that contains the information on the displacement of a point on the earth's surface;
- φ_{atm}: is the contribution to the total interferometric phase due to the path change of the electromagnetic wave induced by the atmosphere;
- φ_{err}: residual phase error.

Referring to Figure 7.24 below, it can be demonstrated that

$$\varphi_{\text{topo}} = \frac{4\pi}{\lambda} \cdot B_{\perp} \cdot \frac{z}{r_1 \sin \vartheta}. \tag{7.20}$$

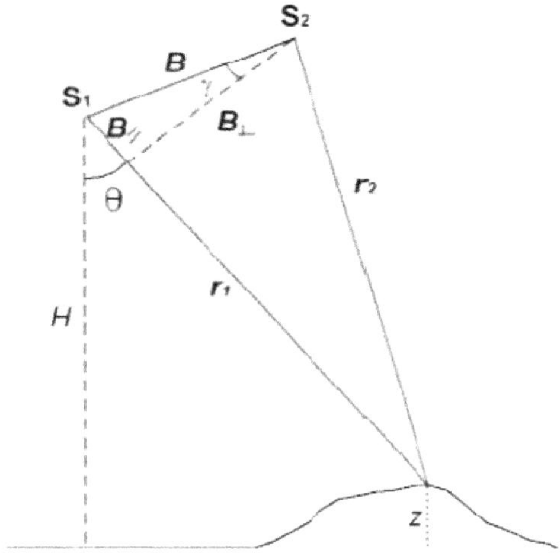

Fig. 7.24 Additional dual mode geometry details.

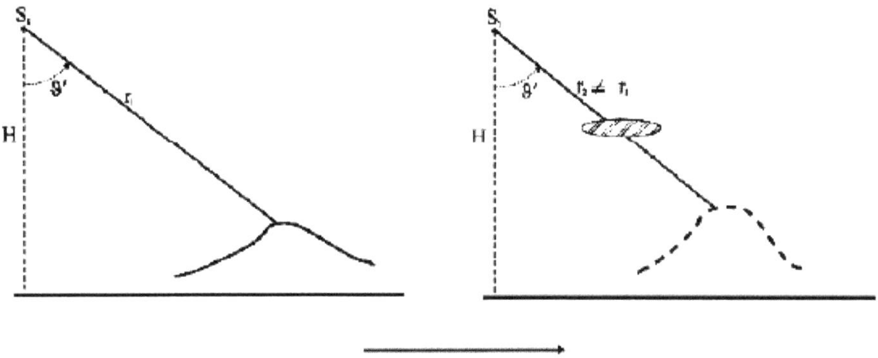

Fig. 7.25 Variation in the length of the path of propagation due to the presence of conditions atmospheric different between the two acquistions.

Thus, the increase of the distance between the two satellites at the data acquisition time, increases the capacity of the InSAR technique to reconstruct the smallest variations in altitude.

Since the acquisition of SAR images by satellite for interferometric uses does not take place simultaneously, in the two different acquisitions are possible changes in the atmospheric refractive index due to different conditions of humidity, temperature and pressure. This leads to a change in the sensor-target distance measurement between the two images, such as to impair the quality of the interferogram (Figures 7.25 and 7.26).

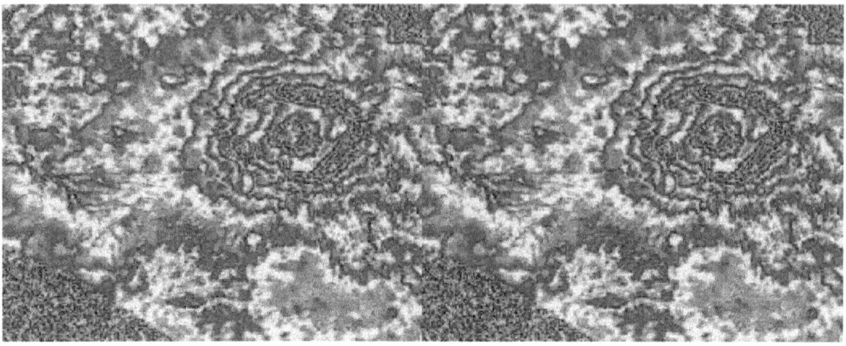

Fig. 7.26 Interferogram with (left) and without (right) atmospheric contribution.

7.4.2 DIF SAR or DIN SAR

It is a variant of the In SAR technique and allows calculating the displacement of the earth's surface. If two satellites S_1 and S_2 pass onto the same area before and after a natural event (earthquake, volcanic eruption, dry tide…) which produces a surface displacement field, as stated above, the interferometric phase φ_{int} will contain also this contribution φ_{displ} due to surface movements (Figure 7.27):

$$\varphi_{\text{displ}} = \frac{4\pi}{\lambda} \cdot \delta r' \tag{7.21}$$

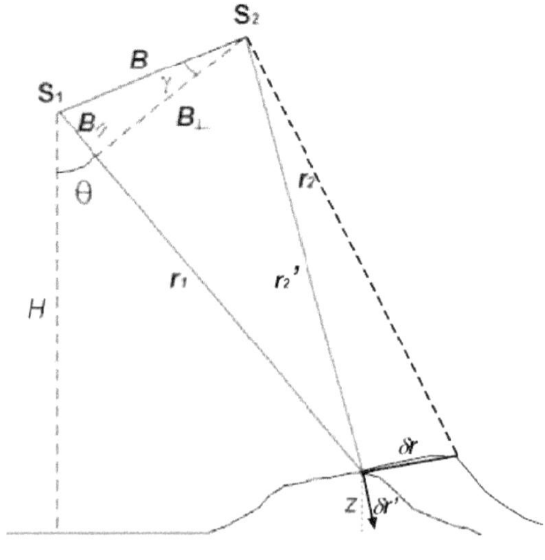

Fig. 7.27 Including surface movements in the dual mode geometry.

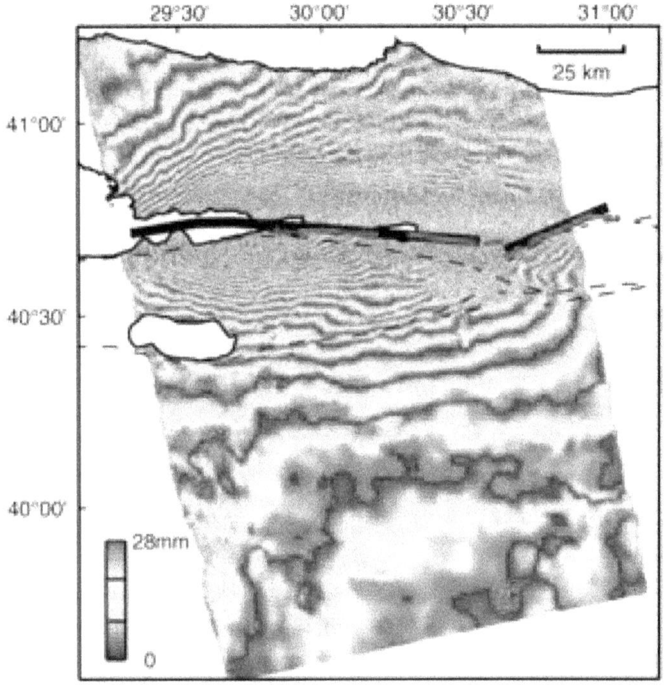

Fig. 7.28 Interferogram (data from ERS 2).

Therefore, we obtain the differential interferogram from the term φ_{int}, after correcting the contribution of flat land and removing the topographic phase, in this case using a DEM. The phase shift will be affected by the errors due to the atmospheric contribution and the residual errors.

Figure 7.28 shows an interferogram, produced using data from the satellite ERS 2, on ground motions produced by the earthquake of Izmir in Turkey (1999):

7.5 The COSMO-Sky Med System

The COSMO-Sky Med program (in the follow on simply COSMO) is the largest Italian investment in the Earth Observation (EO) sector, designed from the outset as "dual use" System for civil and defense uses and so funded and commissioned to the Thales Alenia Space — Italia (TAS-I) consortium by the Italian Space Agency (ASI) and the Italian Ministry of Defense.

Compared to the original concept of privileged coverage of the Mediterranean (COSMO stands for "Constellation of Small Satellites for Mediterranean basin Observation") the system has already shown to have global coverage and capacity to provide data, products and services for applications on a global scale devoted to institutional, commercial, scientific, defense and intelligence sectors, including applications for crisis management and environmental disasters, cartography, hydrology, geology, and the study of marine, coastal, urban, agricultural and forestry areas. The COSMO-Sky Med constellation consists of four satellites in LEO orbit (Low Earth Orbit) to the average altitude of 619.6 km. From June 2010 the system is operating with three satellites, the launch of the fourth satellite took place in November 2010 and the operation of the constellation with four satellites is expected in early 2011. Each COSMO-Sky Med satellite is equipped with a SAR sensor (Synthetic Aperture high-resolution Radar) operating at 9.6 GHz with a wavelength of 3.1 cm, i.e., X-band (8.0 to 12.5 GHz, wavelength between 3.8 and 2.4 cm), and can operate in three different ways — Spotlight, Strip Map, Scan SAR — each characterized by a different combination of resolution and coverage area. Thanks to the SAR sensor and orbital characteristics of the constellation, the COSMO-Sky Med system provides global coverage with high revisit frequency, independently from solar illumination and weather conditions. These features allow the creation of a wide range of applications, from the monitoring of marine and land scenarios up to the investigation of multi-temporal Change Detection images and studies with interferometric coherence analysis.

7.5.1 The COSMO-SKY Med Constellation

The "nominal" orbital characteristics of the COSMO-SkyMed constellation are 4 satellites in the same orbital plane and at 90 degrees from each other (Figure 7.29). This configuration is designed to provide at least four opportunities for shooting any point on Earth the same day, two in Right Looking (default mode) and two in Left Looking mode although under different conditions of observation in terms of angle of incidence.

Since the orbit type LEO — SSO (Low Earth Orbit — Sun-Synchronous Orbit) is almost polar, it offers global coverage and the angle of illumination of the Sun is maintained approximately constant over time; it is typically used

Table 7.1. Orbital characteristics of the Constellation COSMO-Sky Med (SSO: Sun-Synchronous Orbit — LTAN: Local Time of Ascending Node).

Orbit Type	SSO
Inclination	97.86°
Revolutions/day	14.8125
Orbit Cycle	16 days
Eccentricity	0.00118
Argument of Perigee	90°
Semi Major Axis	7003.52 km
Nominal Height	619.6 km
LTAN	6:00 A.M.
Number of Satellites	4
Phasing	90°
Deployment	Progressive

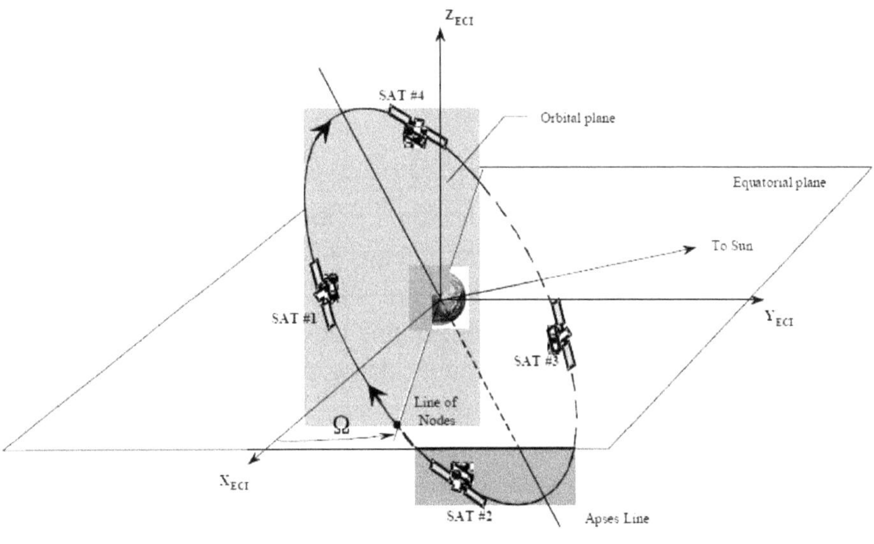

Fig. 7.29 COSMO-Sky Med constellation in nominal configuration with four satellites.

for Earth Observation (EO) applications, for both optical and radar sensors. For COSMO-SkyMed the LEO SSO orbit is almost circular (apogee 624 km, perigee 621 km), “dawn dusk” (LTAN — Local Time of Ascending Node at 6:00 am and pm) and with a rotation period of about 97 minutes. With 14.8125 (14 and 13/16) orbits a day, an orbital period of 16 days consists of 237 orbits.

Fig. 7.30 Swaths of ground cover of COSMO-SkyMed.

The "Ground track Repeatability" of COSMO-Sky Med, a key requirement for Earth Observation, is better than 1 km, i.e., the "sub-satellite point" has a tolerance better than ± 1 km from the nominal ground track: compliance with this requirement is obtained through control of the orbit (Figure 7.30). Furthermore, since the interferometric configuration requires radar observation of a scene under slightly different angles of incidence, the orbital control — beyond the control of the ground track — requires a careful control of the baseline, i.e., the distance between the two points of observation from two satellites, each with its orbit, with an accuracy measured in some tens of meters.

The COSMO-SkyMed constellation operated from 2008 to 2010 with three satellites (FM #1/PFM, FM #2 and FM #3). The fourth satellite, launched in November 2010, is due to become operational in 2011. The three satellite configuration is arranged as depicted in Figure 7.31.

The three satellites, in circular orbit at the nominal altitude of about 620 km, all lie on the same orbital plane, which is tilted at 97.86 degrees; FM #1 and FM #2 are diametrically opposed and FM #3 is located at 67.5 degrees from FM #2, a condition that puts the satellites FM #2 and FM #3 in a "tandem-like — one-day interferometry mission".

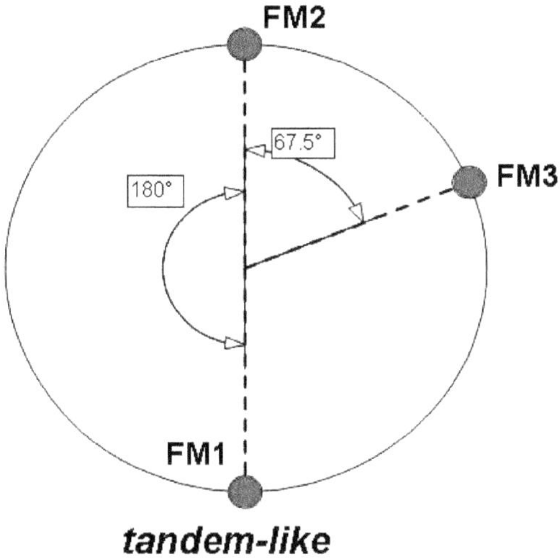

Fig. 7.31 COSMO-Sky Med constellation with three satellites, FM #2 and FM #3 in a "tandem-like — one-day interferometry mission."

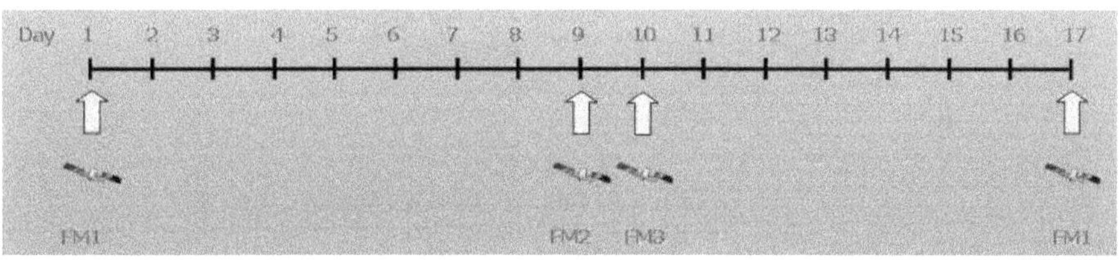

Fig. 7.32 Timing and sequence of the COSMO-SkyMed revisits with three satellites, with FM FM #2 and #3 in a "tandem-like — one-day interferometry."

In temporal terms, since a full orbital cycle of each satellite is 16 days, this means that, taken as a reference FM #1, the conditions of a shooting made by FM #1 — in terms of the geometry of observation, i.e., angle of incidence — recur almost identical for FM #2 after 8 days and for FM #3 after 9 days, then for FM #3 after a single day after FM #2. In terms of application, since the satellites are all in the same orbital plane, they are all in interferometric configuration (IFSAR) as the geometry of observation — i.e., the incidence angle — is repeated; what changes is the de-correlation time: FM #1 and FM#2 are in interferometric configuration at 8 days, FM #2 and FM #3 at one day, FM#3 and FM #1 at 7 days (Figure 7.32).

7.5.2 The SAR Sensor COSMO-Sky Med and Its Acquisition Modes

The SAR sensor — Synthetic Aperture Radar — which is installed on each COSMO-SkyMed satellite operates in X-band (8–12 GHz, a wavelength between 3.75 and 2.5 cm) and in three different modes (Figure 7.33): Spotlight, Strip Map, Scan SAR, each characterized by a different combination of resolution and coverage area.

The satellite flight direction corresponds to the azimuth, the transversal one corresponds to the range.

In all modes of operation of the SAR sensor, the nominal pointing direction of the radar beam with respect to nadir ("look angle" of the SAR) is 34 degrees,

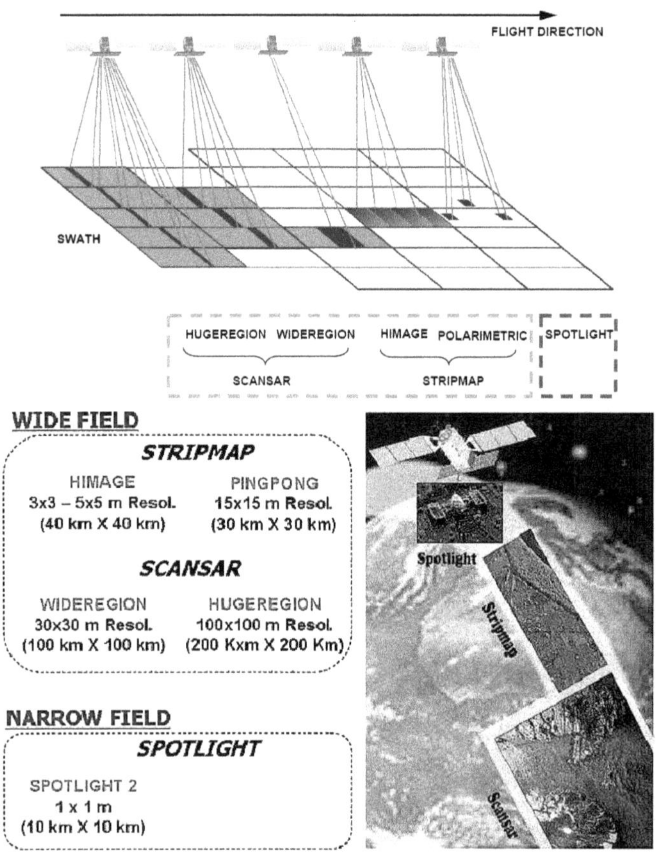

Fig. 7.33 The three modes of acquisition of the COSMO-SkyMed SAR sensor.

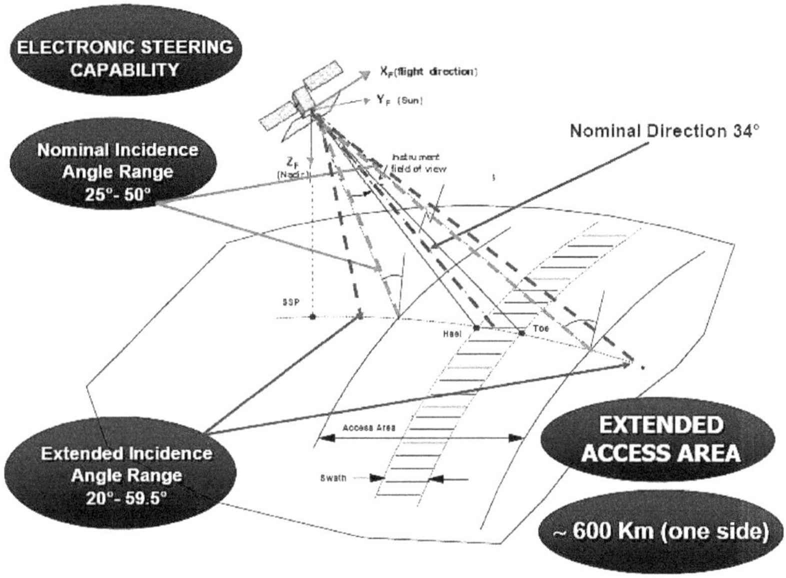

Fig. 7.34 Incidence Angles of the COSMO-SkyMed radar beams.

the nominal field of the angles of incidence of the radar beam is between 25 and 50 degrees and the "extended" one is approximately between 20 and 60 degrees (Figure 7.34).

7.5.3 The Products of the COSMO-Sky Med: Classes and Types

The COSMO-Sky Med system produces three different classes of products (Figure 7.35):

- — SAR standard products: five levels i.e., Level 0, 1A, 1B, 1C, 1D;
- — Higher Level products: processed products, including the interferometric ones;
- — Auxiliary products: products for internal use, like orbital and quality control ones.

The SAR standard products of the COSMO-Sky Med are in accordance with the definitions of international standards for Earth Observation (EO) SAR systems and as such are encoded in standardized "Layers" (Figures 7.36a and 7.36b).

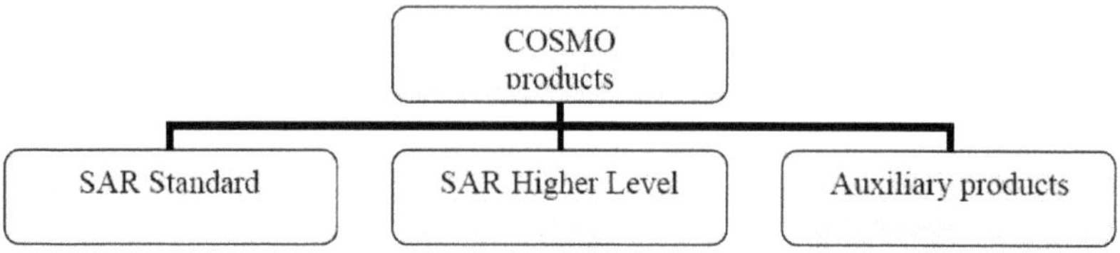

Fig. 7.35 Classes of products of the COSMO-SkyMed.

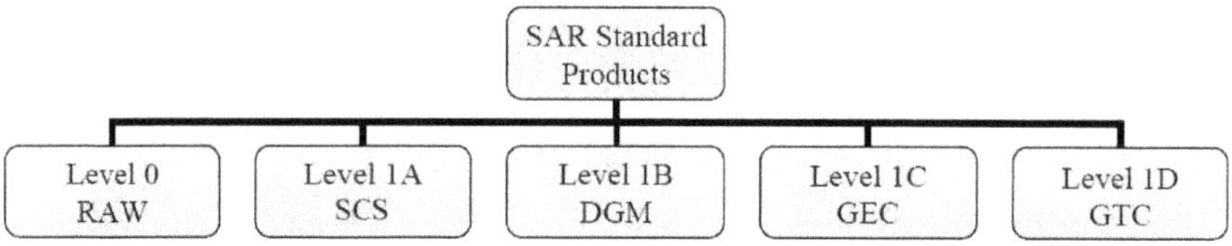

Fig. 7.36a SAR Standard Products from COSMO-Sky Med.

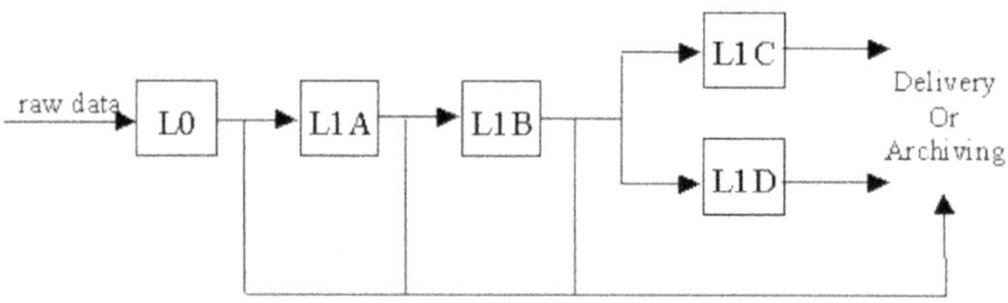

Fig. 7.36b Production flow of SAR Standard products from COSMO-SkyMed.

7.5.3.1 COSMO-SkyMed standard products

The L 0 products are time series of raw data (on board raw data, time-ordered echo data) obtained after decryption, decompression, internal calibration and error compensation, in complex formats with signals in phase and in quadrature (I and Q). These products include all ancillary data needed to develop intermediate products and products of higher levels, such as trajectory, spatial coordinates with accurate date/timing and velocity vector of the satellite, geometric model of the sensor, status of the payload and calibration data (Figure 7.37).

L1A products, otherwise referred to as SCS — Single-look Complex Slant, are raw SAR data in complex format (being the real part the amplitude and the imaginary one the phase) focused with internal radiometric calibration (gain compensation of the receiver and internal calibration) and geometric

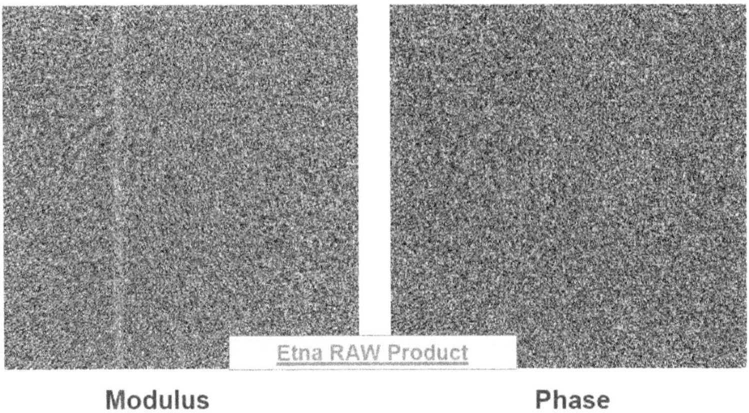

Fig. 7.37 Example of SAR L0 product (E. Lopinto, "COSMO-SkyMed Products and User Services", 2006).

projection zero-doppler slant range-azimuth, i.e., in the projection of natural acquisition of the sensor, with the original spacing geometry and the associated auxiliary data. These products contain focused data in phase and in quadrature (I and Q), weighed and radiometrically equalized.

L1B products, otherwise referred to as DGM — Detected Ground Multi-look, are SAR data in real number (without the complex part) focused, subject to radiometric equalization, with speckle reduced through multi-looking processing (except for those proper to Spotlight mode), processed with detection of the image (width), with geometrical projection zero-doppler ground range-azimuth on a reference ellipsoid or on a DEM — Digital Elevation Model, resampled with a regular spacing on the ground and the associated auxiliary data. The L1B products are obtained from the L1A /SCS — Single-look Complex Slant ones by calculating the amplitude (from complex data), multi-looking (from Single-look, but not for Spotlight products) and projection on a regular grid on the ground (starting from the Slant geometry).

Examples of SAR products L1A and L1B are shown in Figure 7.38.

L1C products, otherwise referred to as GEC — Geocoded Ellipsoid Corrected, are real SAR data focused, computed with detection of the image (width) and despeckled through multi-looking processing, geo-referenced projecting the L1A or the L1B ones on the Earth's reference ellipsoid and with a regular grid obtained from a cartographic reference system, complete with the associated auxiliary data.

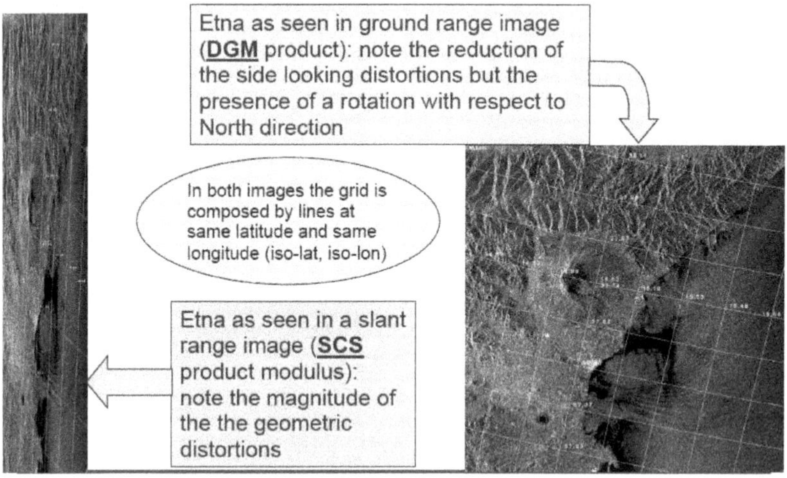

Fig. 7.38 Example of SAR products L1A and L1B (E. Lopinto, "COSMO-Sky Med Products and User Services", 2006).

L1D products, otherwise referred to as GTC — Geocoded Corrected Terrain, are real focused SAR data, computed with detection of the image (width) and de-speckled through multi-looking processing, geo-referenced projecting the L1A or the L1B on a reference DEM — Digital Elevation Model, also with the use of GCPs — Ground Control Points, and with a regular grid obtained from a reference mapping system, complete with the associated auxiliary data.

Examples of SAR products L1C and L1D are shown in Figures 7.39 and 7.40, respectively.

7.5.3.2 COSMO-Sky med — image types

Among SAR Standard products, the COSMO-Sky Med system produces three different types of images corresponding to different modes of operation of the sensor with different combinations of resolution and coverage area (Figure 7.41 and Table 7.2):

— Spotlight, with metric resolution on areas of 10×10 km;
— Strip map — Himage and Ping Pong (polarization HH and VV, or HH and HV, or VV and VH) — with a resolution of a few meters (variable according to product type) over areas of tens of km (40×40 km for Himage and 30×30 km for Ping Pong);

Fig. 7.39 Example of product SAR L1C (E. Lopinto, "COSMO-Sky Med Products and User Services", 2006).

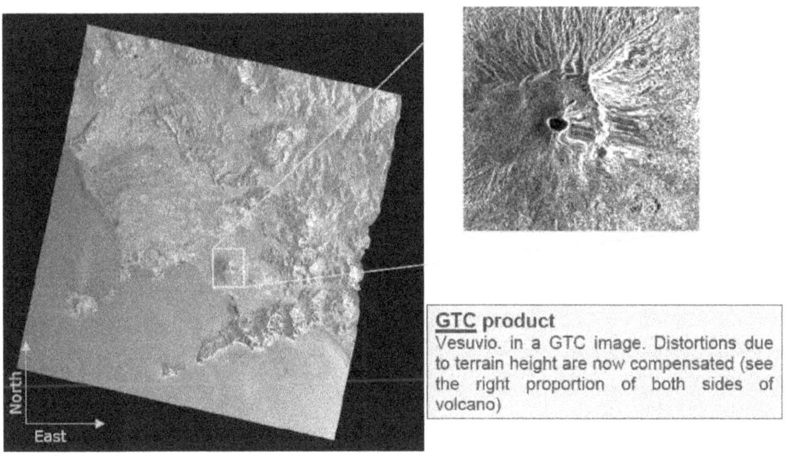

Fig. 7.40 Example of product SAR L1D (E. Lopinto, "COSMO-Sky Med Products and User Services", 2006).

— Scan SAR — Wide Region and Huge Region — with resolutions of tens or hundreds of meters over areas of hundreds of kilometers (100 × 100 km for Wide Region and 200 × 200 km for Huge Region).

In all operating modes, the nominal pointing direction of the radar beam is 34 degrees, the "nominal" field of the angles of incidence of the radar beam

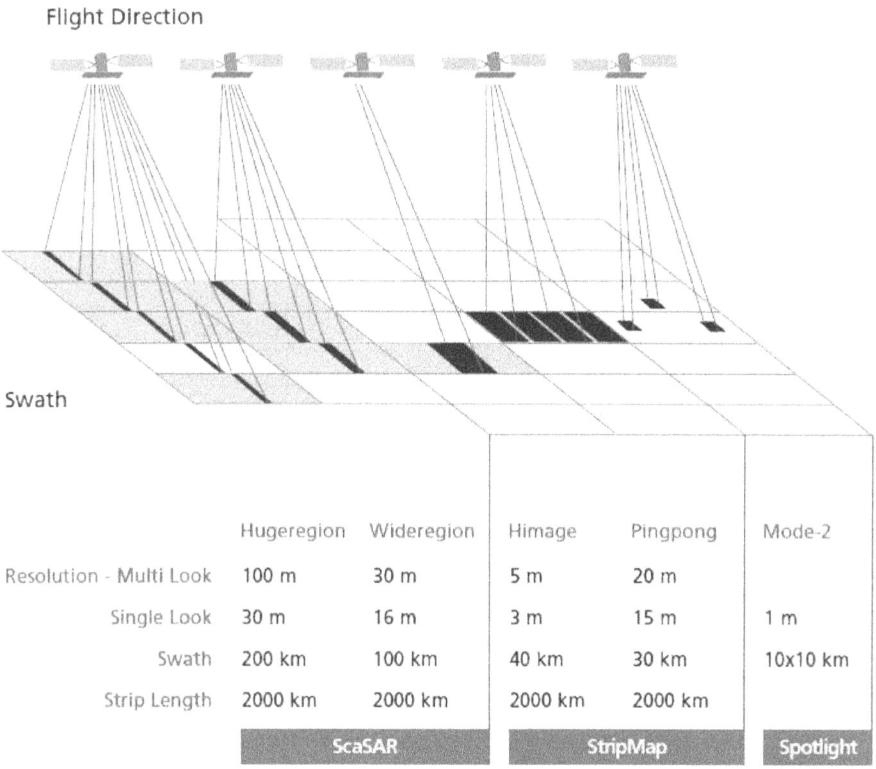

Fig. 7.41 Types of images and operating modes of COSMO-Sky Med (http://www.e-geos.it/products/pdf/e-GEOS_COSMO-SkyMed.pdf).

is between 25 and 50 degrees and the "extended" one is roughly between 20 and about 60 degrees.

7.5.3.3 COSMO-Sky med higher level products

The COSMO-Sky Med SAR Products Handbook contains six types of Higher Level products (Figures 7.42–7.46):

— Quick look: products with reduced spatial resolution, useful for browsing and preview;

— Co-registered: co-registered products, consisting of sets of overlapping images, suitable for investigations of Change Detection and interferometry;

Table 7.2. Operating modes of the sensor, associated types of products and classes of the SAR Standard COSMO-SkyMed products (Levels).

Acquisition Mode	Processing levels:	Level 1A (SCS)	Level 1B (DGM)	Level 1C (GEC)	Level 1D (GTC)
SPOTLIGHT	Line Spacing (m)	0.7	0.5	0.5	0.5
	Pixel Spacing (m)	0.3 ÷ 0.4	0.5	0.5	0.5
	Ground Range resolution (m)	1	1	1	1
	Azimuth resolution (m)	1	1	1	1
	Range × Azimuth scene size (km)	10 × 10			
	Scene duration (sec)	3			
	Geolocation ±3s accuracy (m)		25	25	15
	Number of looks		1	1	1
StripMap HIMAGE	Line Spacing (m)	2 ÷ 0.2	2.5	2.5	2.5
	Pixel Spacing (m)	1 ÷ 2	2.5	2.5	2.5
	Ground Range resolution (m)	3	5	5	5
	Azimuth resolution (m)	3	5	5	5
	Range × Azimuth scene size (km)	40 × 40			
	Scene duration (sec)	7			
	Geolocation ±3s accuracy (m)		25	25	15
	Number of looks		~ 3	~ 3	~ 3
StripMap PINGPONG	Line Spacing (m)	2 ÷ 2.5	10	10	10
	Pixel Spacing (m)	3 ÷ 8	10	10	10
	Ground Range resolution (m)	15	20	20	20
	Azimuth resolution (m)	15	20	20	20
	Range × Azimuth scene size (km)	30 × 30			
	Scene duration (sec)	6			
	Geolocation ±3s accuracy (m)		25	25	20
	Number of looks		~ 2.8	~ 2.8	~ 2.8
ScanSAR WIDE REGION	Line Spacing (m)	10	15	15	15
	Pixel Spacing (m)	2 ÷ 4	15	15	15
	Ground Range resolution (m)	15.0	30.0	30.0	30.0
	Azimuth resolution (m)	27.0	30.0	30.0	30.0
	Range × Azimuth scene size (km)	100 × 100			
	Scene duration (sec)	15			
	Geolocation ±3 s accuracy (m)		30	30	30
	Number of looks		~ 4	~ 4	~ 4

(*Continued*)

Table 7.2. (*Continued*)

Acquisition Mode	Processing levels:	Level 1A (SCS)	Level 1B (DGM)	Level 1C (GEC)	Level 1D (GTC)
ScanSAR HUGE REGION	Line Spacing (m)	15	50	50	50
	Pixel Spacing (m)	7 ÷ 10	50	50	50
	Ground Range resolution (m)	22	100	100	100
	Azimuth resolution (m)	45	100	100	100
	Range × Azimuth scene size (km)	200 × 200			
	Scene duration (sec)	30			
	Geolocation ±3s accuracy (m)		100	100	100
	Number of looks		~ 18	~ 18	~ 18

Source: (http://www.e-goes.it/products/pdf/COSMO-SkyMed_Products.pdf).
For more information please contact: info.cosmo@se-qeos.it

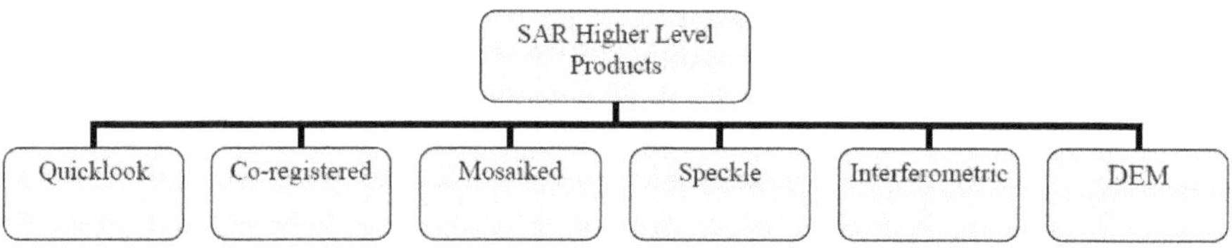

Fig. 7.42 COSMO-Sky Med Higher Level SAR Products.

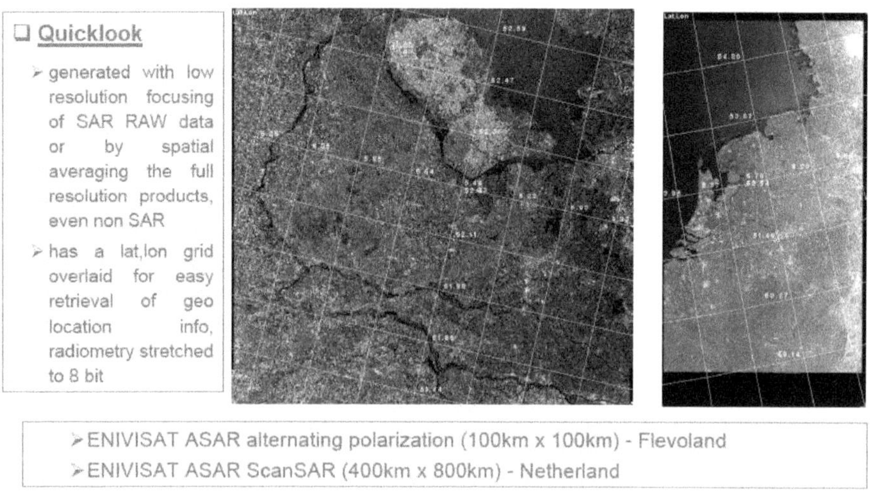

Fig. 7.43 Example of Quic klook product (E. Lopinto, “COSMO-Sky Med Products and User Services”, 2006).

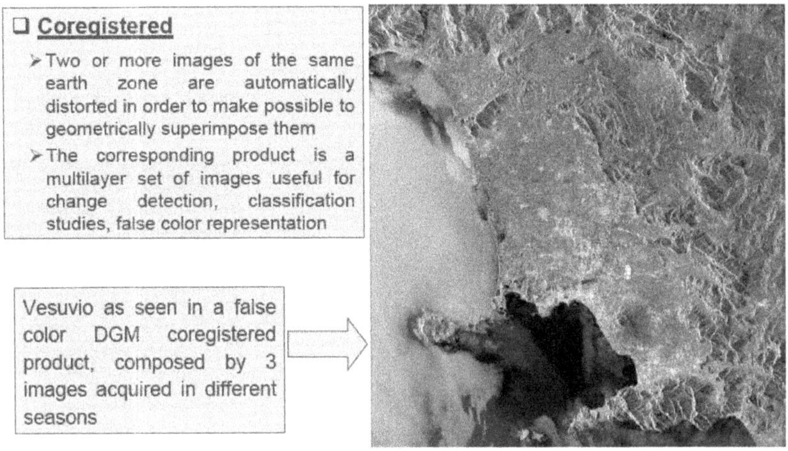

Fig. 7.44 Example of Co-registered product (E. Lopinto, "COSMO-Sky Med Products and User Services", 2006).

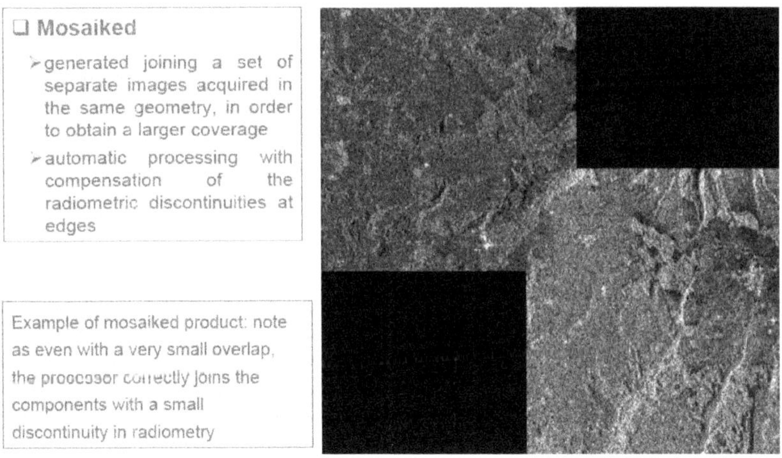

Fig. 7.45 Example of Mosaicked product (E. Lopinto, "COSMO-Sky Med Products and User Services", 2006).

— Mosaicked: "mosaic" products composed of assembled images, typically set side by side with partial overlapping, used for large areas coverage;
— Speckle: filtered images (speckle-filtered) with a higher "equivalent number of looks" (ENL — Equivalent Number of Looks);

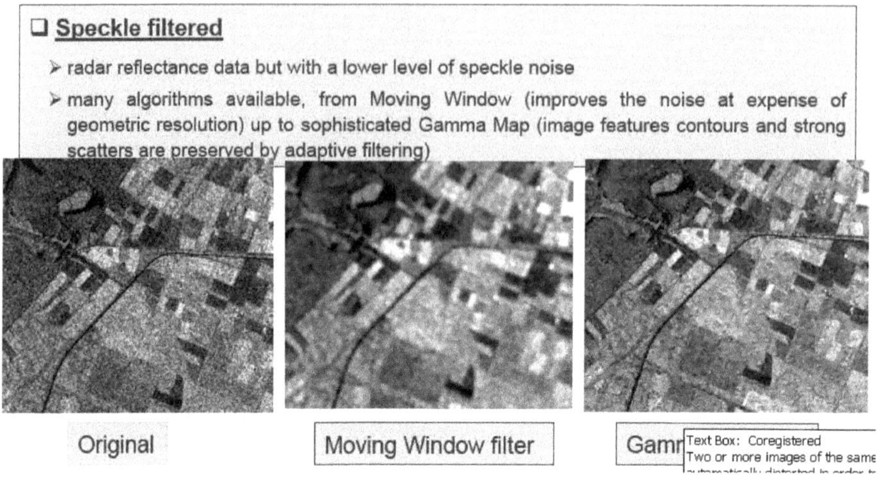

Fig. 7.46 Example of Speckle product (E. Lopinto, "COSMO-Sky Med Products and User Services", 2006).

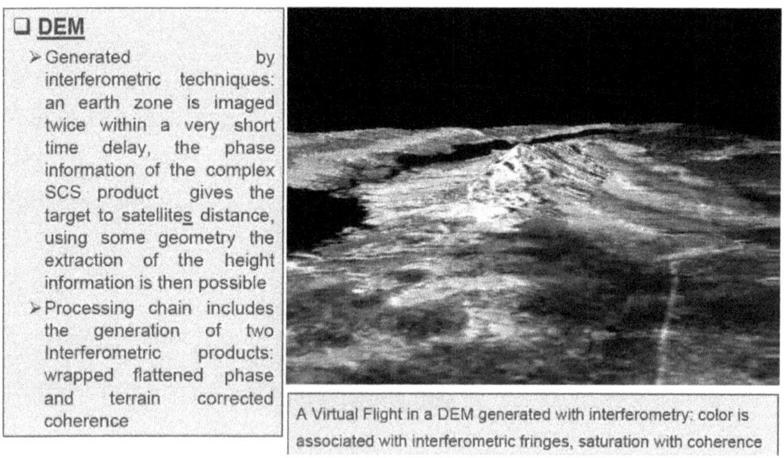

Fig. 7.47 Example of interferometric product — DEM (E. Lopinto, "COSMO-Sky Med Products and User Services", 2006).

— Interferometric: interferometric products, with analysis of coherence and phase, which support a wide range of applications;
— DEM — Digital Elevation Model: digital terrain models which can be obtained with interferometric techniques.

7.6 SAR Applications

The main SAR applications that can be exploited thanks to the elaboration of SAR products can be classified into the following categories:

1. *Mapping*
2. *Damage Assessment*
3. *Monitoring*
4. *Target detection*

7.6.1 Mapping

From a general point of view, among the foremost capabilities of SAR, the acquisition of a large amount of images allows to rapidly populate dedicated data archives, so as to grant a continuous and worldwide mapping. This can be profitably exploited for topographic applications, such as change detection, analysis of areas selected for their high risk of possible emergencies, interferometric stacks (time series) on areas showing possible subsidence phenomena.

In particular, COSMO-SkyMed system provides the capability to focus its data acquisitions over specific areas of interest with the value-added to obtain images within a short time frame thanks to the constellation high performance of its revisit time.

The following image (Figure 7.48) shows the mapping accomplished in about one month over the Italian peninsula using StripMap acquisition mode, executed with a constellation's configuration foreseeing 3 COSMO-SkyMed satellites, of which two in Tandem-like geometry (meaning that the passage over the same area is one day):

7.6.2 Damage Assessment

Particular characteristics of the SAR sensor provide the capability to exploit the data acquired for specific applications. More in detail, the high performance associated to the response time (thanks also to the all-weather and day/night acquisition capability, as well as the satellite's right/left looking features), together with the worldwide access, make SAR systems a foremost systems for damage assessment applications.

Furthermore, the detection of features of interest within the image is improved through the possibility to compare acquisitions over the same scene

Fig. 7.48 Mapping of the Italian peninsula.

at different timeframes, which is possible thanks to the high stability of the instrument's performances (e.g., temporal radiometric stability).

In particular the high geometric resolution and generally speaking the high image quality performance of COSMO-SkyMed SAR sensor, allows a profitable comparison of images acquired at different times, with the aim to highlight changes occurred within the area of interest.

Examples of damage assessment and monitoring are:

- Earthquake
- Flood
- Volcano
- Oil spill

The aforementioned COSMO-SkyMed characteristics allow to evaluate damages occurred after an event, exploiting either a single image (e.g., to assess the extension of a flooding or an oil spill) or multi-temporal acquisitions (e.g., earthquake damages, interferometric applications).

Examples are shown in Figures 7.49–7.52.

7.6.3 Earth Surface Monitoring

COSMO-SkyMed system above mentioned high performances, together with wide swath acquisitions, high image coherency and high resolution features,

Fig. 7.49 Haiti earthquake.

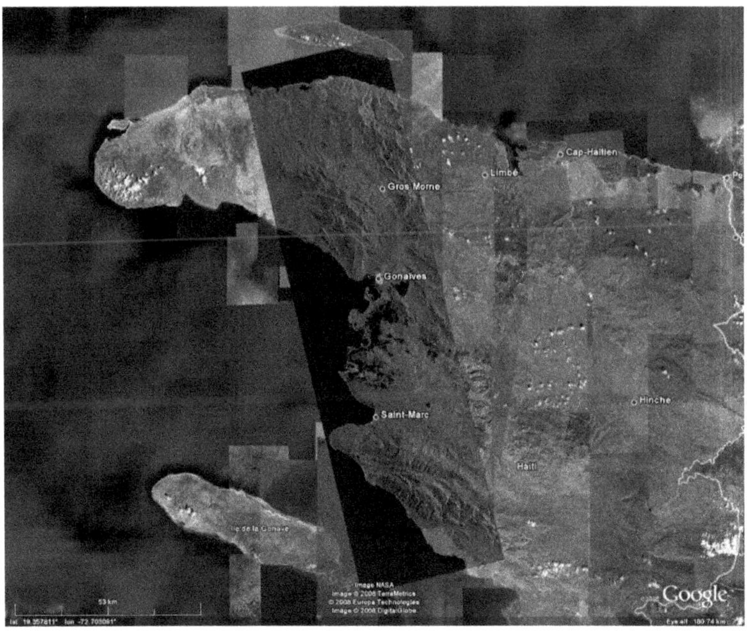

Fig. 7.50 Hurricanes Hanna/Ike (Hanna September 6, 2008; Ike September 7, 2008 COSMO-SkyMed acquisition September 8, 2008 StripMap mode).

Fig. 7.51 Iceland-Eyjafjöll volcano eruption.2010/04/03.Strip Map mode COSMO-SkyMed.

are a value-added for the monitoring of the Earth's surface. Possible applications are:

- Ice monitoring
- Agronomic pattern detection

Examples are shown in Figures 7.53 and 7.54.

Reliable and objective information on cropped area is important to farmers, local and national agencies responsible for crop subsidies and food security, as well as for traders and reinsures.

In the past years is has been shown that SAR based products can provide information on field processing conditions like ploughing, field preparation etc. and on crop growth status such as planting, emerging, flowering, plant maturity, harvest time, and frost conditions. These products — complemented

Fig. 7.52 Louisiana oil spill; COSMO-SkyMed satellite n. 2. 2010 05 03, 23:57.

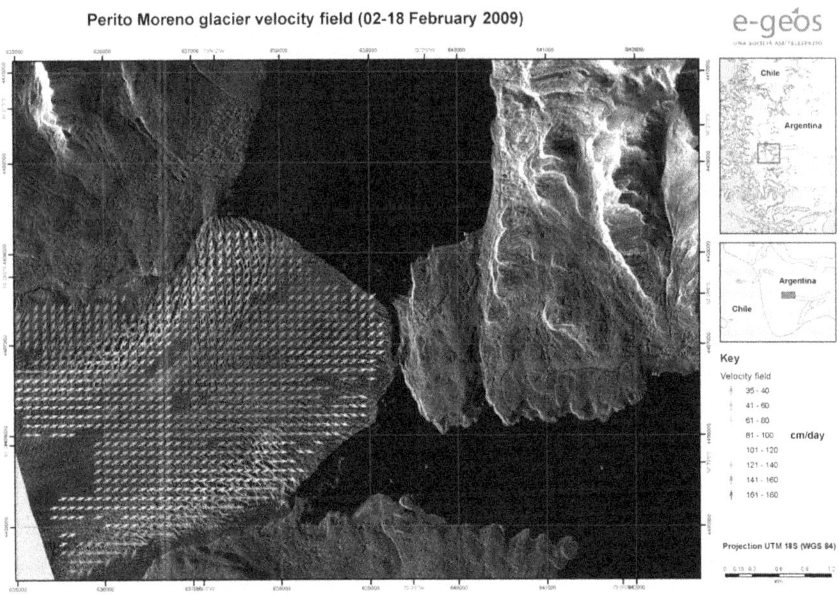

Fig. 7.53 Perito Moreno glacier velocity field.

November (Red) September (Green) Coherence (Blu)

Fig. 7.54 Agronomic pattern detection.

with optical based products (such as chlorophyll, leaf area index, salinity, vegetation indexes, etc.) offers a complete information for a better management of the cropped areas but also of the environment.

7.6.4 Target Detection

This primary application, historically ascribed to SAR systems due to their already mentioned day/night and all-weather acquisition, has gained further importance through COSMO-SkyMed thanks to its higher image quality and revisit time performances. Products stemming from COSMOSkyMed acquisition modes provide microwave images over wide areas (i.e., 10 km $\times$ 10 km swath) reaching metric resolutions, thus emphasizing both materials and shape of the target.

Examples are provided in Figures 7.55 and 7.56.

7.7 SAR and Geomorphology

The monostatic Synthetic Aperture Radar imaging with one antenna for transmission and reception of the backscattered echoes does not bring information concerning the terrain topography: the instrument is performing measures of

Fig. 7.55 High resolution target detection B52 bomber.

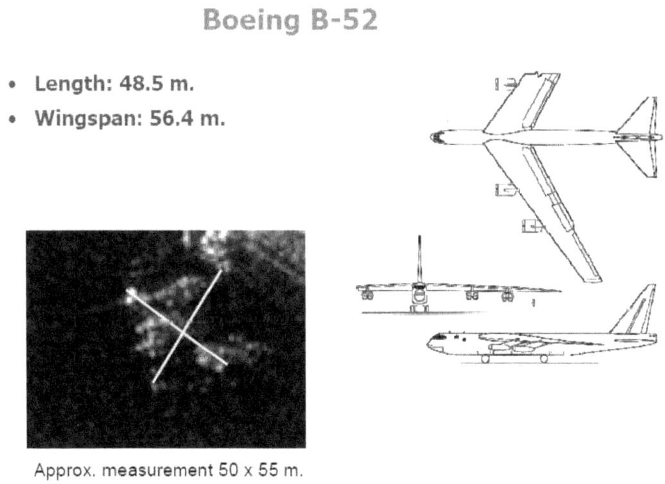

Fig. 7.56 B-52 bomber.

the time delay in the range direction and does not have accurate information about the elevation angle of the backscattered signal which is related to the terrain height.

The usage of two antennas separated in across-track direction, by a *baseline* distance, allows to overcome this limitation and thus to retrieve accurate information about the surface topography. The range distance between the target and the two antennas can be measured exploiting the phase information of the SAR image: the phase difference between the two images acquired by the two antennas is the *interferometric phase* and this principle is at the basis of the SAR interferometry.

The range difference brings information concerning the terrain height for the pixels within the image with an accuracy that results to be far beyond the one achievable with optical space system and stereoscopic techniques. In February 2000 a dedicated mission has been launched with the purpose of generating a global Digital Elevation Model (DEM) of the Earth surface: the Shuttle Radar Topography Mission (SRTM); the Shuttle was equipped with two radars operating in C-band (NASA/JPL project) and X-band (ASI/DLR project). The results consist in the generation of an X-band DEM covering roughly the 40% of the Earth surface with 30 meter of horizontal accuracy and 6 meters of height accuracy and a C-band DEM mapping the 80% of land area with horizontal and vertical accuracy of 100 meters and 12 meters, respectively.

Bistatic SAR operates with distinct antennas for transmission and reception. For SRTM the antennas are mounted respectively one in the cargo bay and the other at the end of a deployable boom (at a distance of 60 m). For space borne systems the antennas are mounted on separate platforms. Bistatic SAR constellations offer a natural way to perform single-pass interferometry in space; to embark the receiver onto a different platform allows the generation of large baseline (thus capable at producing highly accurate DEM). The formation flying ensure the synchronicity of acquisitions but, on the counterpart, a constellation of two satellites in a tandem configuration have the same revisit time of a single satellite. To improve the monitoring capability (in term of response time) the typical approach consists in distributing the constellation in the orbital plane an by a proper phasing to allow the generation of the desired baseline in successive passages (multi-pass interferometry). The absence of simultaneity between the acquisitions in a multi-pass interferometric product introduces the undesired phenomena of temporal decorrelation that lowers the achievable performance in producing a digital elevation model due to the variations within the acquired scene and/or within the travel path (different propagation in the atmosphere).

On the other hand, the temporal evolution of the Earth surface can only be monitored by performing several acquisition in the time domain: to highlight possible surface deformation or subsidence phenomena it is required to analyze historical series on the area of interest. Differential SAR interferometry (DIFSAR) is the remote sensing technique that allows the investigation of such deformations with an accuracy that is a fraction of the radar wavelength (in the range of centimetres to millimetres). Thanks to its capability to produce spatially dense deformation maps this technique is becoming very important in civil protection scenarios. In addition, the monitoring of the temporal evolution of the detected displacements is a key issue that provides information in understanding the dynamics of the deformation phenomena. To this end a time series of the deformation map can be generated, exploiting the available acquisitions, by processing in an appropriate sequence the DIFSAR interferograms.

Since DIFSAR technique is capable at measuring small ground displacement, the main application that can benefit of these capability is the Geology. Monitoring of spatial distribution of surface deformation is a primary goal of Tectonics, which concerns generally with the structures within the lithosphere of the Earth, particularly with the forces and movements of orogenies responsible of earthquakes, and volcanic belts, affecting much of global population. Earthquakes are caused mostly by rupture of geological faults, but they are also induced by volcanic activity, landslides, mine blasts, and nuclear test. At the Earth's surface, tectonic earthquakes manifest themselves by shaking and sometimes displacing the ground. They occur where there is sufficient stored elastic strain energy to drive fracture propagation along a fault plane. Traditional monitoring techniques of such displacements uses levels, total stations and GPS, that can only measure on point-by-point basis, and hence are comparatively costly and time consuming. Through differential synthetic aperture radar interferometry, or Din SAR is possible to gather dense information related to deformation across a large area in an efficient and economic manner, even on dangerous or inaccessible areas. Information obtained by Din SAR techniques can be used to constrain analytical and/or numerical models to derive parameters such as depth, dimension, orientation, and eventual slip occurred on the fault plane, Din SAR techniques allow observing the temporal evolution of the surface displacements, and are thus able to derive the pre-seismic state as well as of the post-seismic effects in an area stroke by an earthquake.

Swelling of the volcano indicates that magma has accumulated near the surface, so it is crucial to measure the tilt of the slope and track changes in the rate of swelling of the volcano sides. In fact, an increased rate of swelling, due to the increased pressure in shallow magma chamber, especially if accompanied by an increase in sulfur dioxide emissions and harmonic tremors, indicates the high probability of an eruption.

Ground subsidence can also be due to underground mining and oil extraction can be also analysed through the Din SAR techniques. The safety of surface infrastructure such as building, roads, bridges, and power lines is crucial to avoid environmental problems such as water pollution, cessation of stream flow et cetera. But there is also a Groundwater-related subsidence of land resulting from groundwater extraction, a major problem in the developing world and major metropolises. The 80% of serious land subsidence problems associated with the excessive extraction of groundwater making it a growing problem throughout the world. Landslides includes a wide range of ground movement, such as rock falls, deep failure of slopes and shallow debris flows, which can occur in different environments. Gravity is the primary driving force for a landslide, but other factors can contribute to affect the slope stability. Pre-conditional factors build up specific sub-surface conditions that make the area/slope prone to failure, whereas the actual landslide often requires a trigger before being released.

Repeat-pass satellite differential interferometry quality is highly dependent on the features of the imaged surfaces (such as topography and vegetation coverage) and on the meteorological conditions. Precisely, these two major limitations in applying DIFSAR are known as spatio-temporal decorrelation and atmospheric artifacts. Generally, the longer the time interval between two acquisitions, the lower is the signal-to-noise ratio of the interferometric phase due to random surface change over time. This may even lead to a failure while trying to detect ground deformations, this happens typically in slowly deforming and heavily vegetated areas. The SNR also decreases with the orbital separation of the two SAR acquisitions due to the different radar viewing angles, thus limiting the number of interferometric pairs that can be used. In addition, atmospheric inhomogeneity can cause varying phase delays that cannot be filtered out by phase differencing and, therefore, can contaminate deformation measurements: *atmospheric artefacts*. These main limitations of DIFSAR technology can be overcome by adopting a multi-interferogram framework.

Several approaches have been conceived in order to mitigate such problems and obtain an accurate time series analysis on the investigation area: the Small-Baseline Subset (SBAS) approach implements an *ad-hoc* combination of the generated DIFSAR interferograms by choosing the interferometric pairs whose spatial baseline results to be small (to minimize the spatial decorrelation effects). The number of acquisition used for the analysis is increased by properly linking independent SAR dataset separated by large baselines. The availability of both spatial and temporal information on the processed data is used to identify and filter out atmospheric phase artefacts; therefore, spatially dense deformation maps and, at the same time, deformation time-series for each investigated pixel of the imaged scene can be produced.

The SBAS technique was originally designed to monitor deformations occurring at a relatively large spatial scale and does not result to be, as it is, appropriate for analyzing local deformations (single buildings or structures). A new algorithm that extends the monitoring capability of the SBAS technique to localize displacements by investigating full-resolution DIFSAR interferograms, still relies on small-baseline interferograms but is implemented by using two different dataset generated at low (multilook data) and full (single-look data) spatial resolution, respectively. The former is used to identify, via the SBAS approach, large-scale deformation patterns, topographic errors in the available digital elevation model (DEM), and possible atmospheric phase artefacts; the latter is investigated after removing the low-resolution signal components; indeed, structures highly coherent over time are identified on the residual phase signals jointly with an estimate of their local topography and of the mean velocity of the residual deformation. A final step, leads to the estimation of the temporal evolution of the nonlinear components of the local displacement affecting these highly coherent structures. The availability of both deformation and topography information, at the two different spatial scales, allows to analyze the deformation behaviour of the investigated pixels and to correctly localize them in a geographic (or cartographic) reference system.

Another technique is based on the Permanent Scatterers (PS) concept to discriminate the displacement from topographic phase and atmospheric signature. The PS are privileged point-wise radar targets that are only slightly affected by decorrelation: typically correspond to man-made structures (or part of them) in urban areas and /or bare rocks in rural area. Not a single PS is necessary to implement this technique, a sparse grid of such privileged points is necessary to estimate and remove the atmospheric signature. The estimate of both elevation and deformation can be faced on a pixel basis in a long temporal series of SAR data acquired over the area of interest along the same (nominal) satellite orbit. All available images are focused and co-registered on the grid of the unique master acquisition. The master image have to be selected as a baseline barycentre to keep a low value the dispersion of the normal baseline value. Afterwards the amplitude stability index is computed (on the radiometrically equalized images) to infer information concerning the expected phase stability of the scattering barycentre of each sampling cell. PS candidates have to show a high stability index. If a DEM is available, the topographic phase can be removed (at least its most part); successive processing steps estimate the residual topographic phase term. The phase of a generic pixel within all the interferograms can be thought as the sum of four different contributions: (a) the possible target motion (w.r.t. its position in the master image), (b) atmospheric contribution, (c) decorrelation noise and (d) the residual topographic phase contribution. PS approach aims at separating (and resolve) these phase terms in the time–space–acquisition geometry multidimensional analysis: target motion (a) is correlated in time, can exhibit different degrees of spatial correlation and is uncorrelated to the baseline. Residual topographic phase (d) is proportional to the baseline, results uncorrelated in time but can be differently correlated in space depending on the reference DEM (if any). The atmospheric contribution (b) is strongly correlated in space and uncorrelated in time and baseline. The first step is the estimate of Atmospheric Phase Screens (APS). Thanks to the high spatial correlation of APS, even a sparse grid of PSC enables the atmospheric components retrieval. The elevation estimate that can be subsequently carried out at PSC couples can be regarded as a multi-baseline phase unwrapping on a sparse grid. The fundamental is the joint exploitation of large and small baseline values to increase reliability and accuracy of elevation estimate (high sensitivity provided by large baseline with the reduced risk of introduction of unwrapping errors thanks to the small baseline).

A novel approach, named Persistent Scatterers Pairs (PSP), have been proposed to identify and analyse the persistent scatterers in a series of full resolution SAR images. The atmospheric effect is filtered out by exploiting its spatial correlation without relying on models. The PSP method allows to efficiently identify the persistent scatterers, and to retrieve their terrain height and displacement velocity. Extensively used on real data, it results to be effective in generating high density of persistent scatterers with particular emphasis in the urban areas.

The permanent (or persistent)-scatterers (PS) IFSAR technique was proposed to extract deformation signals from a set of interferograms and to estimate the atmospheric phase screen and digital elevation model (DEM) errors. The PS-IF SAR technique models and analyzes PS targets that maintain high SNR in phases. PS-IF SAR technique uses a single master image. Several extensions to the PS-IF SAR approach have been proposed, e.g., by connecting the PS points with the Delaunay triangulation method: a technique that is a compromise between the PS and SBAS paradigms since it detects PS based on coherence instead of amplitude threshold. The number of required images can be reduced but, the spatial resolution of the PS points is limited, and it is impossible to detect certain good PS candidates surrounded by decorrelated pixels. Another connection method derived from PS-InSAR approach for extracting

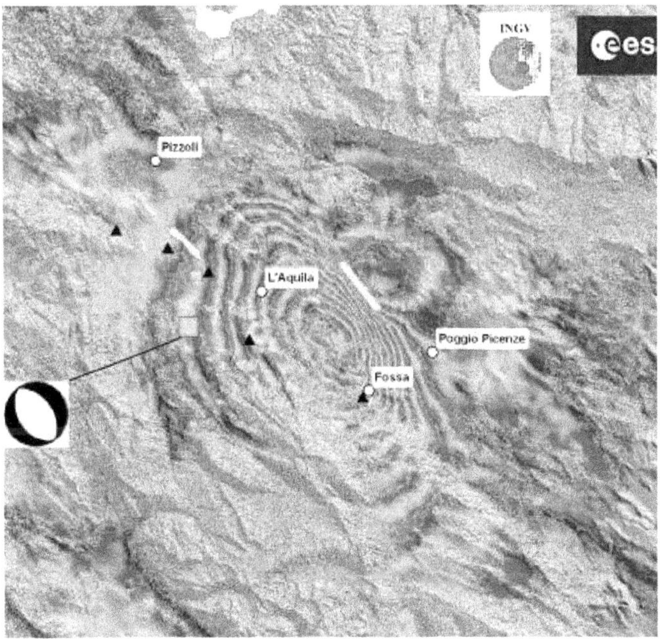

Fig. 7.57 ENVISAT Interferogram: Each fringe corresponds to about 2.8 cm of displacement in the satellite direction (23° from vertical).

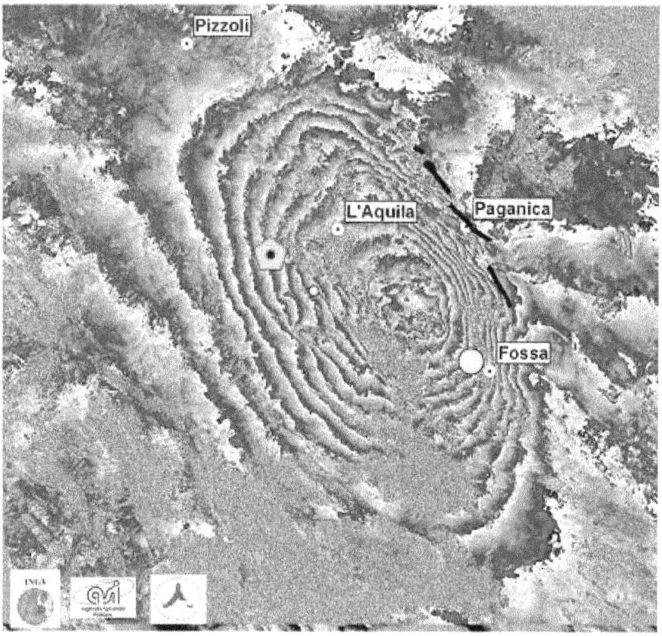

Fig. 7.58 COSMO-SkyMed Interferogram: Each fringe corresponds to about 1.5 cm of displacement in the satellite direction (36° from vertical).

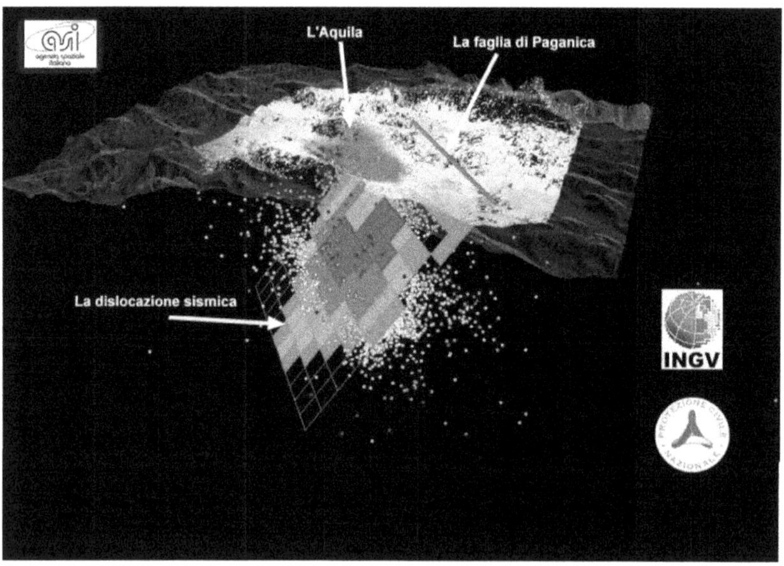

Fig. 7.59 Maximum seismic dislocation along the fault (red) corresponds to maximum ground deformation (25 cm) measured by COSMO-SkyMed satellites.

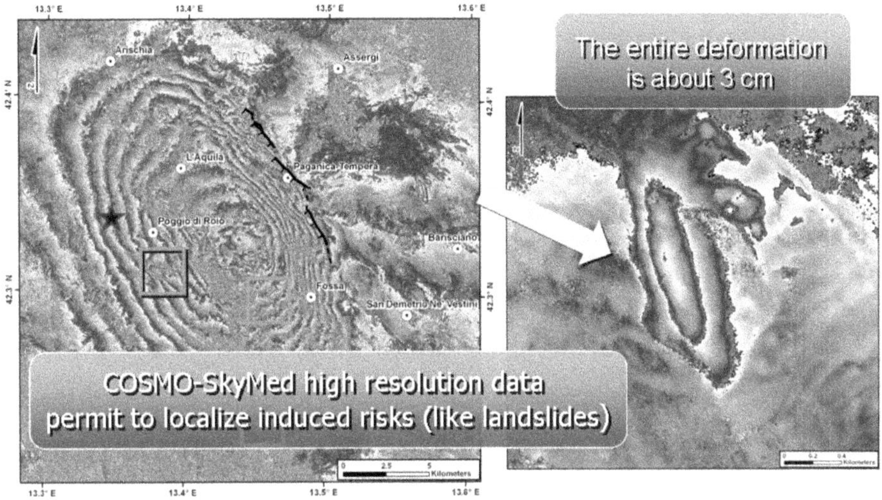

High resolution interferograms allow the detection of small deformations. The green box in the upper figure shows an area that is subject to a continuous deformation, the process have been speed up by the earthquake (Possible landslide in the future).

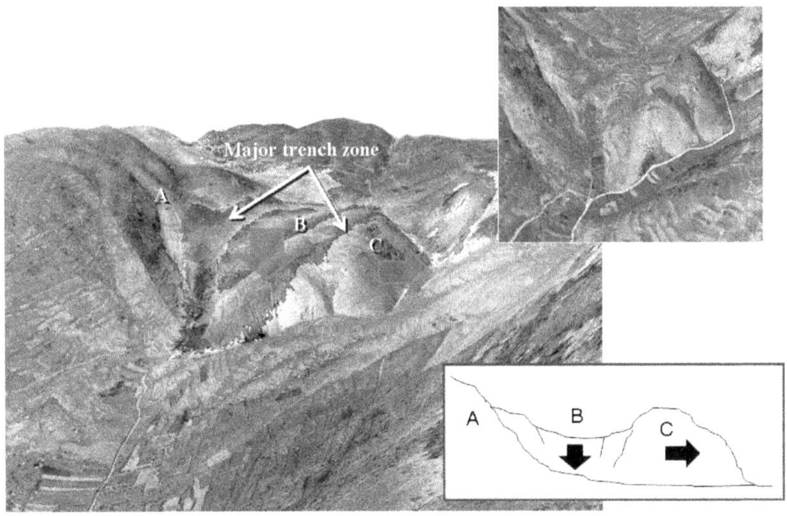

Fig. 7.60 High resolution interferograms..

deformation signals from differential interferograms proposes to connect all PS points within a given threshold of distance, forming a so-called freely connected network (FCN). Only interferometric combinations that have both small spatial and temporal baselines are used in the solution. The linear defor-

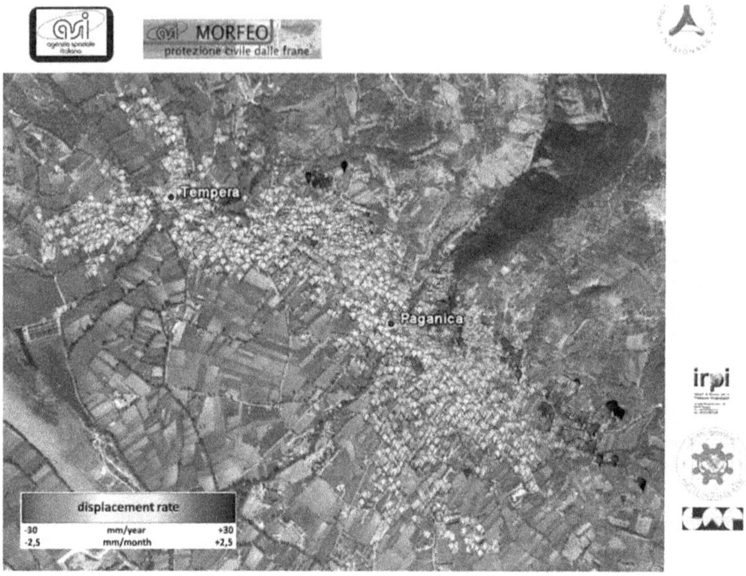

Fig. 7.61 High resolution map — single points displacements close to the Paganica fault — 26 COSMO-SkyMed StripMap images processed (acquisition time range 12 April-20 September).

mation velocities and DEM errors are estimated based on the LS principle and after removing observation outliers. The time series of phase measurements is reconstructed by SVD (singular value decomposition) as done in the SBAS technique and then decomposed into nonlinear deformations and atmospheric signals by using the empirical mode decomposition.

Examples are provided in Figures 7.57 and 7.58.

The fault plane obtained from the computed model starting from COSMO-SkyMed data is shown in Figure 7.59 The fracture plane has a dip of about 50° towards the SW and passes under L'Aquila city. During the earthquake, the Earth crust block located SW from the fault plane slided downside for a maximum slip of 90 cm at 4 km-depth, producing the ground subsidence pattern shown in Figures 7.60 and 7.61 by red colour.

References

[1] I. Merril. Skolnik Radar Hanbook 3rd edition(2008) Mac Graw Hill.

[2] G. Franceschetti and R. Lanari, *Synthetic Aperture Radar Processing*, CRC Press, 1999, ISBN 0-8493-7899-0.

[3] *COSMO-SkyMed System Description & User Guide*, Italian Space Agency, COSMO SkyMed Mission, rev. 6 agosto 2010 — URL: http://www.e-geos.it/products/cosmo.html. e URL: http://www.asi.it/it/flash/osservare/cosmoskymed.
[4] *COSMO-SkyMed SAR Products Handbook*, Italian Space Agency, COSMO-SkyMed Mission, rev. 6 agosto 2010 — URL: http://www.e-geos.it/products/cosmo.html e URL: http://www.asi.it/it/flash/osservare/cosmoskymed.
[5] http://www.e-geos.it/products/pdf/e-GEOS_COSMO-SkyMed.pdf (Sistema).
[6] http://www.e-geos.it/products/pdf/COSMO-SkyMed_Products.pdf (Prodotti).
[7] http://directory.eoportal.org/get_announce.php?an_id=8990 (COSMO-SkyMed).
[8] E. Lopinto, "COSMO-SkyMed Products and User Services", 2006, URL: http://www.oosa.unvienna.org/pdf/pres/stsc2006/ind-07.pdf.
[9] J. R. Wertz and W. J. Larson (eds.), *Space Mission Analysis and Design*, 3rd ed., Kluwer Academic Publishers, 1999, ISBN 1-881883-10-8.
[10] J. R. Wertz, *Coverage, Responsiveness, and Accessibility for Various "Responsive Orbits"*, 3rd Responsive Space Conference, April 25–28, 2005, Los Angeles, CA, AIAA RS3-2005-2002.
[11] J. C. Curlander and R. N. McDonough, *Synthetic Aperture Radar Systems and Signal Processing*, Wiley, 1991, ISBN 0-471-85770-X.
[12] *Exploring ENVI*, ITT Visual Information Solutions, 2008, edizione fuori commercio.
[13] *Synthetic Aperture Radar and SARscape — SAR Guidebook* — sarmap, September 2008.
[14] Ruggero Stanglini, Anacleto Mocci, *COSMO-SkyMed — The Italian Space System for Earth Observation*, Agenzia Spaziale Italiana — Segretariato Generale della Difesa e Direzione Nazionale degli Armamenti, 2008, edizione fuori commercio.
[15] P. Finocchio and R. Prasad, Marina Ruggieri (eds.), *Aerospace Technologies and Applications for Dual Use — A New World of Defense and Commercial in 21st Century Security*, River Publishers, 2008, ISBN 978-87-92329-04-2.
[16] *Using Earth Observation Data for offshore exploration*, in *PetroMin — Asia's Exploration and Production Business magazine*, September 1999, AP Energy Business Publications Pte Ltd, Singapore, ISSN 0129-1122.
[17] *Monitoring Oil Production Facilities Using Radar Coherent Change Detection*, in *GEO Informatics — Magazine for Surveying, Mapping and GIS Professionals*, vol. 13, July/August 2010, CMedia crossmedial publisher, ISSN 13870858.
[18] *Golfo nero — I più grandi giacimenti petroliferi USA scoperti negli ultimi decenni si trovano negli abissi del Golfo del Messico*, in *National Geographic Italia*, ottobre 2010.
[19] http://www.asi.it.
[20] http://www.e-geos.it.
[21] http://www.cstars.miami.edu (Università di Miami).
[22] http://www.telespazio.com.
[23] http://www.thalesaleniaspace.com.
[24] http://www.ittvis.com (ENVI).
[25] http://www.creaso.com (SARscape).
[26] http://rapidfire.sci.gsfc.nasa.gov/realtime (MODIS — Moderate Resolution Imaging Spectroradiometer — Rapid Response System).
[27] http://www.goes.noaa.gov (GOES — Geostationary Operational Environmental Satellite).
[28] http://www.jaxa.jp (GMS — Japan's GeostationaryMeteorological Satellite).

[29] http://earth.esa.int/resources/softwaretools.
[30] http://www.oosa.unvienna.org (United Nations Office for Outer Space Affairs).
[31] http://www.worldenergyoutlook.org (World Energy Outlook).
[32] http://www.ft.com/cms/s/0/c68f728a-935b-11df-bb9a-00144feab49a.html (Petrobras).
[33] http://www.npagroup.com (Oil and Gas).
[34] http://fugro.com (Oil and Gas).
[35] http://www.centricenergy.com (Oil and Gas).
[36] http://www.esa.int/SPECIALS/Operations/index.html (Space Situational Awareness).
[37] http://www.mna.it (Kamil Crater).
[38] http://www3.lastampa.it/scienza/sezioni/news/articolo/lstp/284792/ (Kamil Crater).
[39] http://www.nationalgeographic.it/scienza/2010/07/27/news/fresh_crater_found_in_egypt_changes_impact_risk_-78402/ (Kamil Crater).
[40] http://www.focus.it/Community/cs/blogs/una_finestra_sulluniverso/archive/2010/07/23/407993.aspx (Kamil Crater).
[41] http://www.isis-online.org (Institute for Science and International Security).

Part III

The Experimental Network

8

A Swarm Sensor Network for Earthquakes

Francesco Fedi*, Sabino Cacucci*, Simone Barbera†, Ernestina Cianca†, Marina Ruggieri†, Cosimo Stallo†, Massimo Guerriero‡, and Giovanni Ottavi‡

*Space Software Italia
†University of Roma Tor Vergata — CTIF_Italy
‡ITP

8.1 AIM

Humans face natural hazards at different scales in time and space. The hazards affect life and health: they have a dramatic impact on the sustainability of society, especially in societies that are vulnerable because of their geographic location, poverty, or both. the first decade of the 21st century has been characterised by a lot of natural disasters, such as earthquakes (e.g., Sumatra-Andaman in 2004, Kashmir in 2005, Sichuan in 2008, L'Aquila in 2009) followed by landslides (China in 2008) or tsunami (Indian Ocean in 2004); floods (e.g., West and Central Europe in 2002; China in 2007; Taiwan and Philippines in 2009); cyclones (e.g., Katrina in 2005; Nargis in 2008) and several others. On 12 January 2010 Haiti was struck by a violent earthquake; its epicenter was located nearby the capital city Port-au-Prince. According to Haitian government, more than 111,000 people died in the earthquake, 194,000 were injured and 609,000 became homeless [1].

According to this vision, it is mandatory to study and develop very efficient tools in order to forecast these natural disasters that have caused in many areas of the world economic and social damages. In particular, the United Nations, with the support of the European Community, are now trying to develop initiatives finalized to the prediction and to the early warning of such events [2].

Currently, the Italian scientific community is trying to spread the seismic risk awareness to the population through the so-called "Seismic Risk Chart." From this work it came to life the "National Institute for Geophysics and Volcanology" (INGV), and the operative branch is the "Italian Civil Protection." Despite of this, the country is not yet well organized, in terms of systematic procedures and advanced technologies.

A pool of companies and academic and research centers got recently in touch and agreed to start a cooperation finalized to the solving of this kind of issues. A first analysis of the current situation and of the future ideas has been done during the seminar "Seismic and Hydro Geologic Events. Make Italy Safe!" held in Rome on November 2009. The main peculiarity of this pool of companies and academic and research centers is to be heterogeneous, in terms of expertise, covering different areas as telecommunications, geology, electronics.

A project has been proposed, based on the common idea that the earthquakes prediction is feasible and that plenty of data can be handled and broadcast within the needed time and to the proper institutions. The overall architecture is based on the "Top-Down Logic", inducing the spontaneous development of virtuous behaviors of the local communities without any dictation coming from the high level of the chain. Finally, the project wants to be a support to the existing institutions, and it does not want to be a replacement.

8.2 Why A SSN for Earthquakes

This paragraph describes the primary mission of the SSN within a system for the Management of Earthquake Risks.

During the second International Congress on prevision of earthquakes, which was held in Lisbon on April 2009, it was asserted that: "Institutions and Not Governmental Organizations (NGOs) interested in the prediction of earthquakes should define test areas by using independent experts and financing research activities in this field." Finally it was concluded that "current efforts in rescue and rehabilitation activities are very expensive and could be more effectively and conveniently moved toward prevention actions."

The problem of forecasting in general and in real-time of the behaviour of seismic wave always represents one of the most important research issues.

Throughout the history of seismology, the issue of earthquake prediction has been characterised by conflicting opinions. These difficulties mainly arise from the lack of a common understanding on how a prediction should be defined. According to the suggestion of the IASPEI Sub-Commission for Earthquake Prediction, an earthquake precursor is defined as "a quantitatively measurable change in an environmental parameter that occurs before mainshocks, and that is thought to be linked to the preparation process for this mainshock" [3]. The set of ideas that lead to the quantitative definition of a precursor is recognisable as a "hypothesis", or a "model". It should be stressed that the hypothesis, or the model, characterising the concerned anomaly or precursor, has to be defined in a univocal way, so that it could be objectively recognised and evaluated in any circumstance and by any observer. In [4] the authors have defined the steps of the process leading to the acceptance of a given precursor by the application of a rigorous statistical approach. These steps include a phase during which the parameters of the model are estimated by the analysis of past cases (*learning phase*), and a successive phase where the hypothesis is tested against a new and independent data set (*testing phase*). Non-random time-space distribution of seismicity can be recognised itself as a precursory phenomenon of forthcoming earthquakes.

In seismology, two different aspects of non-random occurrence are commonly taken into account: earthquake clustering and long-term quasi-periodicity. The first kind of behaviour is frequently observed as a short or medium term interaction among earthquakes, which tends to occur in groups, such as foreshock-mainshock-aftershock sequences, multiplets and swarms. The second kind of behaviour is supposed to characterise strong earthquakes that appear separated by fairly regular time intervals, when they occur on the same seismic sources. Both these non-random behaviours, although they appear of contrary nature, can be properly modeled and used for earthquake forecasting.

Different kind of mathematical models have been developed in order to describe the process of earthquake event. Among them there are the stochastic models and hybrid ones, respectively.

The main scope of stochastic modeling consists of defining the three phases of the earthquake wave and identifying the main parameters for each phase, such as resonance frequency, damping ratio and peak value. The stochastic models are based on the hypothesis that every wave has a transient character since its source is active only for a short time and its energy will be damped and finally absorbed by the body of the earth [5]. Key parameters derived from current empirical seismic hazard analyses, as response spectra, peak ground acceleration and power spectra are very often based on one-dimensional extracts from recorded or simulated three-dimensional seismogram data, which are crisp values. A nonlinear dependence among these parameters has been noticed at each site from earthquake to earthquake. According to a classic earthquake stochastic model [6], each wave is dividing into three phases and it shows different spreading velocity of longitudinal or primary (P-waves) and transversal or secondary (S-waves), causing an S-wave of the same origin to arrive later than the corresponding P-wave. The third phase (C-/G- waves) is connected with converted and guided waves. Moreover, it has been noticed that waves during an earthquake have more than one origin. While the main source of all is the earthquake fault, emitting so called direct waves, sudden changes of material density cause reflections as well as P-S, S-P and other wave conversions anywhere along the propagation path. Waves of different type and origin usually differ in the share of energy and the frequency content, and overlay each other. The first phase is connected with domination of direct P-waves, which have earliest origin and highest velocity. With increasing the distance from epicenter, the first arrival of S-waves is delayed, and with its arrival, the direct S-wave dominates the principal direction since its share of energy is usually the greatest of all wave types. When the phase of direct P-waves has finished, the characteristics of the S-wave are very clear. As the energy of the direct S-wave decays, indirect waves converted at discontinuities of layered rock determine the state, alone or together with surface waves. It has been discovered that one of the most important parameters of an earthquake event is represented by the predominant frequency at the first P-phase of the earthquake excitation and her comparison with the predominant frequency at the second S-phase. The value of this parameter is fundamental to forecast the behaviour of seismic waves.

The most important differences for identification of all types of body and surface waves are represented by the characteristics of particle movement. Particles are moved in parallel to the spreading direction by longitudinal waves, but perpendicularly to the transversal waves of either polarization. Surface waves roll particles within a plane parallel or perpendicularly to the surface plane. If the share of energy of one type of wave is significantly larger than the others, it dominates the adjustment of the principal plane or direction of acceleration at this time. When this dominance is strong enough over a certain period, the principal direction should remain at a certain state. If there are high changes of share in the present mix of waves, there will be significant changes in the accelerograms.

The artificial intelligence methods are usually adopted in order to forecast the resonance frequency of S-wave on the base of registrated resonance frequency of P-wave. Neural networks are trained with real seismic records [7] in order to forecast the behaviour of the earthquake process.

Other models are based on the fact that crisp values as earthquake parameters can be successfully characterised with the help of fuzzy logic models [8].

Currently, one of the very promising trends consists of creating models, which combine different approaches like neuro-fuzzy one and stochastic and artificial intelligent ones. Such hybrid models use the machine learning capabilities of neural networks and combine it with transparency and representation power of fuzzy logic and stochastic models. The combination of neural networks and fuzzy logic leads to improve the real time performance, robustness and accuracy of certain control system. This approach can be adopted for real-time forecasting of strong motion acceleration according to general, tectonic, seismic and site parameters with the aim of realising a reliable system for structural control. On the basis of the results forecasting, a right decision for activating different devices for passive, active or hybrid structural control for protection of high risk structures like power plants (in particular, nuclear power plants) can be applied.

Therefore, in order to realise a very reliable system, it is clear as "intelligent models" should integrate heterogeneous information coming from many different analyses that:

- have outlined several correlations between some observed events and the disaster;
- have discovered time and spatial correlation among Earthquakes by historically studying the region. The previously introduced correlations have been observed in the main seismic regions of the

Earth (Japan, Anatolia, California). As result of this analysis it could be possible to predict the risk of seismic events: in an area where an event is expected (and energy has been collected) the days with the highest probability of seismic energy release could be identified.

- have pointed out how the seismic processes are predominantly activated in the time periods of maximum gravitational interaction inside the Earth-Sun-Moon system. These results have been achieved through the analysis of earthquakes information from the United States Geological Survey (USGS): half of the registered seismic events have been registered within two days from the maximum gravitational interaction phase. This correlation is strictly linked to the Earth rotation and Moon tides cycles and could give us precious information to understand earthquakes generation mechanism.

Nevertheless, these heterogeneous sources of information are not always collected and when collected, there is no effort so far in correlating all of them. Moreover, new technologies are nowadays available also to collect real-time data and distribute them.

The realisation of a Seismic Disaster Ahead Management Swarm Sensor Network (SD SSN) could represent the most suitable solution in order to provide a short-term predictability model for Earthquakes which will allow authorities to improve the "response" through prompt action and definition of priorities. This objective will be pursued by collecting and properly processing information coming from heterogeneous sources and apparently "uncorrelated" actors: data from natural processes (collected from the Earth or from satellite) by also data from local people after a disaster and observed behavior of animals just before the disaster.

Therefore, the novelty provided by a SD SSN is mainly represented by:

- a new approach in collecting information;
- a novel approach in the type of information that is gathered and correlated;
- new methods to process it Web-based dissemination and distribution platform.

8.3 The SD-SSN Architecture

The main aim of a SD-SSN architecture is to promote cooperation between the scientific community, the local companies and the local community, to implement a flexible tool finalized to mitigate the risks. This tool should take into account the following factors:

1. The chemical and physical variations of the territory are monitored by a lot of sensors, conceived for different purposes, but *integrable* with the seismic monitoring instruments in the most dangerous areas.
2. New semantic instruments can allow collecting information from data that, otherwise, would be hardly measured and poorly organized.
3. The currently collected data contain "rules", not yet identified, which can bring to the mathematical description of the seismic events; an apparently "quasi-random" event, as the earthquake, should actually be well described by time-space equations.

To implement this toll, the use of several individuated initiatives, along with data currently available, can help to trace the actions to be performed before and after the earthquake. The innovation consists both in the idea itself and in the scientific methods; in particular, the use of the "swarm intelligence" is a huge factor of originality and efficiency.

The useful data can be categorized as:

- Real-Time Data:
  a. Local Data:
     1. From Earth: swarms of sensors over sample buildings which should release micro movements and torsions;
     2. From the sky: GPS and SAR data for monitoring risk areas.

  b. Global Data:
     3. From Earth: local and global seismic and geological data;
     4. From the sky: local and global meteorological data.

- Survey-Based Data:
    a. Interviewing affected people;
    b. Data collected in call centers per relieving abnormal events, data collected from first responders and people after a disaster (interviews).

The main tools used for data processing are:

- Artificial adaptive systems, as artificial neural networks, used to manage extremely complex information in order to predict earthquakes and to create maps of seismic risk factors.
- Semantic analysis for heterogeneous information analysis; in particular to understand, extract and code useful data in a database and to identify hidden relations between different data.

The data collection and the exchange of information are easily performed through the development of swarm network architectures, of wireless networks and of advanced sensors. The territory security requires new paradigms in order to design decentralized and cooperative complex systems answering in an adaptive manner to the scenario evolution. If the information is widespread along the territory, the decision quality improves and the answer time is drastically reduced. The use of the "swarm intelligence" means using autonomous individuals, able to cooperate with others and to be adaptive to the environmental variations. In the earthquake context, a swarm network can help in phase of prediction and in phase of first rescue. The swarm logic should be applied to a sensors/actuators network, able in this way to self-organize in answer to a seismic event.

8.3.1 SIAP

The *Adaptive Informative System for Prevention and First Rescue* (SIAP, Italian Acronym) provides an infrastructure able to drive the proper information, concerning a seismic event, towards people in the proper places and times. The aim is to support the operator to take the best decision. This is an example of autonomy, achieved through the use of "swarm intelligence" techniques.

SIAP is an autonomous network, made up of multi-sensors able to distribute any information towards the users. These data are selective, provided only to the enabled users, which are the users having that right in that moment.

Such distribution is based on high information reliability, availability, promptness and security. The SIAP reticular nature allows achieving a complete vision of the seismic phenomena, through the installation of one or more SIAP nodes on each single building to be checked. The information value is increased from the ability to cooperate among the nodes providing time and space correlated data, with the further advantage to individuate possible patterns describing the event under-observation.

The density and adaptive ability of the nodes, along the working capability even without any server, makes the SIAP system intrinsically robust, able to provide the service even with some malfunctioning nodes; it is exploited, in this sense, the swarm intelligence. Each single node is equipped by sensors providing seismic and structural information, referring to the place where it is installed. Such information can be used both by mobile and fixed operators. Information is geo-referenced, in order to give a detailed and updated picture of the territory.

Each node can drive some switch, acting autonomously to prevent further damages occurring in case of pipes or power wires breaking.

The single node can also manage lighting and emergency systems; through the cooperation among the nodes lighting systems can automatically activate themselves, with an important advantage for the rescue operators.

SIAP can also be an ad-hoc communication network, to exchange data among different operative nodes, or can be a database collecting what happened in a building before a catastrophic event.

Finally, there is an interface between the SIAP and a *Risk Analyzer*, in order to "make intelligent" the collected information.

8.4 The SD-SSN Services

The main services of the SD-SSN are based on the capacity to achieve a prediction and an early warning in case of earthquakes. This can be done through the integration of different areas, involving geologists, engineers, mathematicians and so on. The result of this combined effort is to provide to the citizen the following main services:

- Continuous Monitoring of the Buildings Status.
- Ability to Predict a Seismic Event with a High Probability.
- Efficient First Rescue.

A fundamental coordination service must be provided by institutional organizations. SD-SSN can, in fact, make available a great amount of data, but all these data must be exploited in the proper way to guarantee a serious coordination and synergy between Civil Protection, Police and Fire Brigades in the operative phases.

8.5 Exploitation of SSN Services: The Situational Awareness

ITP Elettronica has accumulated thirty years of experience in the development of original software, for client-server and web-oriented multiplatform environments, aimed for dealing with issues of territorial impact simulation of phenomena, with extensive use of techniques of 2D/3D representation on the territory of surface distribution variables, obtained by interpolation of data produced by sensors, and intersecting with other types of spatial information (e.g., population density).

The development of mapping functions and GIS capabilities, the production of prevision/simulation and command and control systems (in civil and military environments) complete the field of produced applications.

Adaptive Information System for Prevention and early intervention (SIAP) may use the software made available by ITP Elettronica to make more

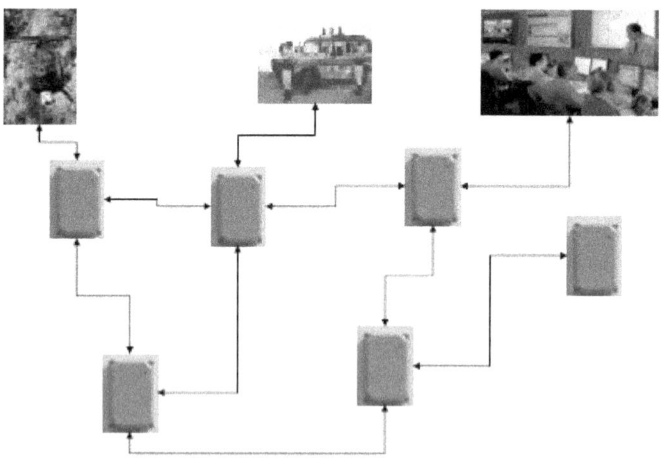

Fig. 8.1 SIAP network.

Fig. 8.2 SIAP nodes on a building.

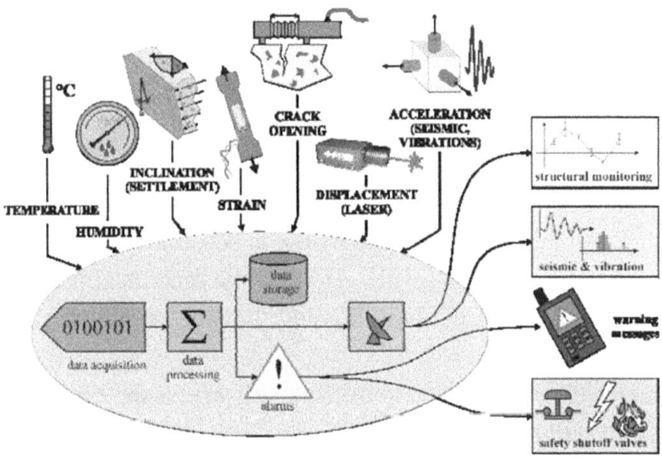

Fig. 8.3 SIAP sensors and actuators.

immediate and easier to interpret the information generated and distributed by the system itself, in order to simplify and make more effective the decision process of the involved operators. The ability to see the actual location of each node of the multisensor network allows an immediate perception of the action area of each sensor and of the level of coverage of the surveillance zone to which one is interested.

Having the appropriate satellite maps and/or photogram metric aerial survey (the only ones that give enough detail), makes it possible in the first place

Fig. 8.4

to get the real 2D representation (2D plan) of the area of interest, as shown in Figure 8.4. On this georeferenced map it is possible to place, in their exact locations, the network nodes of the SIAP. If the map used has enough resolution you can get to see the building and the side on which each sensor is located.

This perspective allows you to immediately assess the correct distribution of sensors that will monitor the entire area. The software can then be queried using the mouse, to obtain detailed information of the single sensor, the latest values of physical measures detected (and properly sent by wireless links), the status information of the sensor and of the possible present actuators, the situation of the connections to other network nodes and collection points. These are all elements that can be shown directly and provide a powerful tool for monitoring and detailed analysis of the situation in the scenario.

The network ability to rearrange the interconnections between the nodes themselves and the centers of data collection even with the fault of one or more connections (re-routing of information) can be effectively presented on the territory by lines of interconnection between network elements, as shown in Figure 8.5. It is therefore straightforward to assess the critical situations

Fig. 8.5

that may lead to the isolation of part of the surveillance area and take all the necessary actions to solve the critical issues in place.

Even more realistic and more significant in some ways are the representations in which the third dimension is taken into account: 3D representations.

Given a proper resolution orographic model (usually a height database in which the values are organized as a regular matrix or as properly shaped and dimensioned non-omogenic triangle vertexes) it is possible to approximate earth surface in the area of interest. By means of 3D graphic techniques (such as texturing) it is therefore possible to color such surface by spreading on it a properly georeferenced satellite or aerial image. The result, managed by a specific software, can be considered a real model of the territory in which it is possible to move around to freely observe the resulting scenario from whatever position and whatever direction. This way the relieves, (hills, valleys, plains) turn out to be evident, allowing for an immediate perception of the morphology of the represented area. A further step consists of the more or less accurate reconstruction of the buildings on location. Even though an extremely accurate reconstruction is possible, it would require such a great volume of data regarding the building shapes, the architectonic elements and their position, that it would be too costly to achieve. First of all because of the difficult availability of such data at a proper level of detail (a correct acquisition implies the use of complex hardware such as LIDAR, a high precision ranging laser coupled with a digital camera, that makes possible to acquire, by means of a software elaboration, very detailed 3D models).

In addition a great level of detail implies the synthesis of a great number of polygons making significantly heavier the task of the 3D modeling and navigation software. Luckily, in most cases, only a simplified reconstruction of the buildings is required, in which only the external dimensions of the building itself are taken into account. In such case it is only necessary to know the measures and the shape of the geo-referenced perimeter polygons and the height of the building in order to generate, (using a technique called extrusion) a box that returns the overall volume. By coloring such parallelepipeds with appropriate colors or using the same image used for texturing the terrain it is possible to achieve an acceptable and meaningful result highly indicative of the town of interest, as shown in Figure 8.6.

Like in the 2D map, also in the 3D scene thus obtained it is possible to add sensor representations and their connections to the SIAP network. To highlight

Fig. 8.6

the positions of sensors regardless of what's your point of view within the scene (navigation), the sensors themselves are represented by icons always facing the viewer and above the buildings, which could conceal them from view, and connected to their real location through a seam line. In this way access to all the additional information, which occurs as the 2D view by querying with the mouse, is simplified.

With appropriate interpolation of the individual measures of physical quantities made available to the SIAP network nodes, it is possible to reconstruct the evolution of the same variables extended to the entire area. By linking the values of a magnitude to the intensity of a predefined color is possible to show its variations in space and time, as depicted in Figure 8.7.

Any localized problems and their influence spread to the surrounding areas are therefore of easy and immediate interpretation. The operator can then use this tool to decide, for instance, the actions on the available actuators and verify in real time the effectiveness of the measures taken on the performance of the monitored quantities, or manage any alarms (in automatic or manual) reaching a set threshold. Additional information may be obtained by intersection with other available spatial variables. For example the calculation of the population

Fig. 8.7

included in the area is defined by some variable exceeding a given threshold. Regardless of the type of representation (2D or 3D), specific tools are available to the operator to measure distances (plan metric and/or orthodromic), areas, elevation profiles, ground slopes and evaluation of altitude points.

Analysis tools can then evaluate the variables involved for individual points along lines or inside areas (variation along roads, profiles, isovalue lines, histograms, etc).

Algorithms for optical sight calculation and representation, synthesis of shortest paths from graphs and simulation engines (such as those for the prediction of electromagnetic coverage, diffusion of pollutants, flat areas, etc.) complete the set of available tools.

Finally the use of web technologies allows, in presence of an intranet/internet connection, to use the obtained information simply through a standard browser without the installation of applications and/or plugins on the user's computer.

8.6 Next Steps: The Network Centric System Paradigm and Biological Approach for Long-Term Earthquake Prediction

This paragraph firstly describes a possible exploitation of the Network-centric Paradigm to the whole system for the long-term forecast and subsequent management of the earthquake risks. Finally, it investigates the possibility to exploit the capability of animals to exhibit unusual behaviour in conditions of very small environment anomalies with aim of achieving useful data to insert, together with other heterogeneous source of information, in intelligent models for a long-term earthquake prediction.

Long-term earthquake prediction has been a subject of great interest of the scientific community, aiming to predict forthcoming seismic events a few years to even decades before their actual occurrences [9]. Most applications in that direction focus upon earthquake recurrence times [10, 11] and possible long-term earthquake precursors, such as changes in seismic activity patterns that can occur during the preparation stages of large earthquakes [12, 13]. The in depth-investigation of earthquake catalogues that record the seismic activity at the observed area has enabled scientists to identify several patterns of changes in seismic activity that were followed by catastrophic earthquakes. Despite some sporadic cases of successful real-time predictions [14] and attempts to support possible identified long-term earthquake precursors according to mechanical models [15, 16], such approaches, which are based on the interpretation of long-term earthquake precursors, show the drawback that the observed changes in seismic activity vary considerably from one area to the next [17, 18].

In order to face the uncertainty of the physics describing the earthquake generation and the lack of causal relationships among seismic patterns and related crustal environments, the application of the network-centric paradigm to a SD SSN infrastructure becomes a crucial step in order to achieve a not only short term prediction, but, above all, a long term forecast aimed at reducing the impact of disasters and to produce planning tools for disaster risk mitigation at all scales.

As already illustrated in Chapter 6, the network-centric approach moves the focus of the system analysis and design from the system functions to its organisation. According to this vision, the system is defined as an integrated whole, whose properties directly stem from the relationships among its components. Such a system is itself part of the environment where it acts and interacts with, via a feedback loop where each system action modifies the environment, whose changes impact the next actions of the system. Moreover, as described in Chapter 6, another specific feature of the network-centric paradigm specifically relates with the hierarchical organisation of the systems, the enaction of system properties, i.e., properties which emerge and then called emergent property. In the ecological paradigm, the network is the recursive scheme to look at the complex systems which are studied as network of networked entities which implies organisation of organised entities. In this scenario, the SI bio-inspired approach in a network-centric paradigm could again represent a useful tool in order to obtain and collect new data to insert in an intelligent model for the long-term earthquake prediction.

However, research into the ability of animals to predict large seismic events such as earthquakes has been hampered by the rarity and unpredictability of such events. Earthquakes, unlike other natural hazards such as hurricanes and volcanoes, occur without any reliable preceding phenomena. This precludes the design of experiments to test hypotheses concerning unusual animal behavior in relation to large seismic events, and most such observations were recollected once the earthquake had already occurred.

Much unusual behaviour shown by domestic animals (such as dogs, chickens, cows, etc. that are normally in close proximity to human settlements) occurs shortly before an earthquake event, often coinciding with P-waves, which arrive a few seconds before the damaging S-waves that can be felt by humans [19]. This response to P-waves cannot be defined a predictive response, but rather an early warning system [20]. Behaviour occurring several days or weeks in advance of an earthquake event is rarer. Fish, rodents, wolves and snakes reportedly exhibited strange behaviour up to two months before the Tangshan, China earthquake (28/7/76, M = 7.8) and a month before the Haicheng, China event (4/2/75, M = 7.3), but most unusual behaviour occurred within a day or two of the event [19]. Out of 36 earthquakes occurring between 1923 and 1978 in Asia, Europe and the Americas, most unusual animal behaviour occurred near the epicenter within 1 or 2 days of the earthquake and species primarily reported were domestic. Fish, rodents and snakes were the only animals that showed unusual behaviour more than a week before the event, or at some distances (greater than 50 Km) from the epicenter [19]. Gathering a random variety of physical and chemical parameters near all possible earthquake epicenters is very difficult and expensive, therefore non geological sources, as for instance wild animal behaviour, might be exploited as further data in a network centric system to make long-term earthquake predictions.

There are several possible mechanisms by which the prediction of seismic events by animals may occur. Some animals may be able to detect P-waves (which travel faster through the Earth's than the subsequent damaging S-waves), earthquake lights (anomalous aerial luminosity) or ground tilt, all of which occur seconds to minutes before earthquakes [20].

Groundwater anomalies, increases in humidity and changes in electrical activity may also be detected [20].

The process of humidity reception in animals is known as hygroreception. Spiders and insects have hygrosensitive sensilla which consist of specialized receptor cells with hygroscopic hair-like structures that detect humidity and/or temperature fluctuations [21]. Vertebrates appear to detect humidity through their olfactory system. Controlled laboratory experiments have demonstrated that desert rodents are able to detect seed caches buried in dry sand based on variations of only a few percent of their water contents [22].

Animal detection of impending earthquakes through hygroreception might therefore be possible in arid environments. However, it very difficult to see how this method would work in rainy areas like Japan, which have uniformly high levels of humidity both in soil and in air. It is also challenging to understand how the pattern of a pre-seismic humidity change would differ from that generated by an impending storm. On the other hand, it was noted that some of the behaviours exhibited by animals before earthquakes resemble their pre-storm ones, so this may be a component of their pre-seismic behaviour. Digital hygrometers are on many commercially-available home humidifying systems, and ought to be inexpensive to put in the field.

Geomagnetic anomalies may be detected by animals that have a well-developed magnetoreception system for circadian and navigation purposes. It was discovered that the magnetotactic bacteria use for orientation small crystals of ferromagnetic mineral (Fe_3O_4), which are formed biochemically [23]. These bacteria are held together in linear chains and hence their individual magnetic moments sum together. The resulting magnetostatic orientation energy per cell typically exceeds thermal noise by factors between 10 and several thousand. Further studies have also revealed similar magnetite crystals in honey bees [24], pigeons [25] and fish [26]. It is well known that birds are disoriented by magnetic anomalies. Moreover, the latter cause the cetacean stranding events along the coastlines [27]. Most of positional variance observed in fin whales migrating at sea is well explained by their avoidance of high magnetic fields and field gradients [28]. This suggests that they use marine magnetic lineation as one of the elements of their navigation system. Analyses of these data have demonstrated values of sensitivity to intensity fluctuations of a few nT.

Moreover, it is well known that nocturnal animals and those which live in nest or dark cavities (like honeybees) are not always able to set their internal circadian clocks with sunlight. Hence, they use the diurnal variations in the geomagnetic field as a timing cue. It was proved in that bees raised in a constant-condition flight room were able to maintain track of their internal biological clocks, despite the absence of visual, thermal, humidity and other signals relating to day/night cycles [29]. In [30], it has proved that honeybees have a magneto-reception system that is tuned for maximum sensitivities below 10 Hz. There are some reports where an unusual swarming behaviour of bees has been observed until fifteen minutes before the onset of strong earthquakes.

Moreover, it has proved that animals might be able to detect raised radon gas levels. There are many reports of these anomalies in groundwater before an earthquake event.

Finally, some recent studies have demonstrated that there is link between ionosphere perturbations with large earthquakes. Very Low Frequency (VLF) and Low Frequency (LF) electromagnetic signals can be used to detect ionospheric perturbations caused by seismicity. Superimposed epoch analysis has established that the ionosphere is disturbed a few days to a week before strong earthquakes ($M > 6$) [31]. Moreover, it has been noticed that shallow earthquakes disturb the ionosphere to a much greater extent than ones that are deeper (>30 km) [32]. In [32] it has been proven that toads at San Ruffino Lake (74 km from L'Aquila, Italy) exhibited a dramatic change in behaviour five days before the strong earthquake ($M = 6.3$) occurred on 6/4/2009 at 01:32:39 GMT. They did not spawn and did not resume normal behaviour until some days after the event. The reduced breeding activity exhibited by toads has been linked to pre-seismic perturbations in the ionosphere detected by VLF radio sounding [33].

Therefore, the unusual behaviour exhibited by some animals before an onset of a large earthquake event, as demonstrated by many studies above reported, could be more investigated and analysed. This will aim at achieving very useful data that, together with other information coming from other intelligent sources, in a network centric system scenario, could give very successful results for a long term earthquake prediction to produce planning tools for disaster risk reduction at all scales.

References

[1] http://www.cnn.com/2010/WORLD/americas/01/23/haiti.earthquake/index html.

[2] http://www.saferproject.net.

[3] M. Wyss, Special issue on earthquake prediction, *IASPEI Symposium* held during the 25th General Assembly in Istanbul, Turkey, Tectonophysics, vol. 193, 1991.

[4] D. A. Rhoades and F. F. Evison, "The VAN earthquake predictions", *Geophysical Research Letters*, vol. 23, No. 11, pp. 1375–1378, 1996.

[5] R. J. Scherer and M. Zsohar, "Probability of the SH wave resonance frequency of a random layer over half space," *the first Joint EU-Japan Workshop on Seismic Risk*, Chania, Crete, Greece, March, 24–26 1998.

[6] R. J. Scherer and M. Zsohar, "Stochastic wave propagation methods as an important part of seismic hazard analysis," *the first Joint EU-Japan Workshop on Seismic Risk*, Chania, Crete, Greece, March, 24–26 1998.

[7] S. T. Radeva, "Forecasting the behavior of seismic waves on the base of fuzzy logic models", Proceedings of Automatics and Informatics 2002, Sofia, vol. 1, pp. 101–104.

[8] S. T. Vassileva, "Predicting earthquake ground motion descriptions through artificial neural network for testing the constructions", Structural Engineering, Mechanics and Computation, vol. 2, Elsevier, 2001, pp. 927–934.

[9] R. E. Wallace, J. F. Davis and K. C. Mc Nally, "Terms for expressing earthquakes potential, prediction, and probability", *Bull. Seismol. Soc. Amer.*, vol. 74, No. 5, pp. 1819–1825, October 1984.

[10] E. G. Trip and L. R. Sykes, "Frequency of occurrence of moderate to great earthquakes in intracontinental regions; implications for changes in stress, earthquake prediction, and hazard assessment", *J. Geophys. Res.*, vol. 102, no. B5, pp. 9923–9948, 1997.

[11] R. E. Habermann, "Precursory seismic quiescence: Past, present, and future", *Pure Appl. Geophys.*, vol. 126, no. 2–4, pp. 279–318, June 1988.

[12] M. Eneva and Y. Ben-Zion, "Techniques and parameters to analyze seismicity patterns associated with large earthquakes", *J. Geophys. Res.*, vol. 102, no. B8, pp. 17785–17795, 1997.

[13] B. Shaw et al, "Patterns of seismic activity preceding large earthquakes", *J. Geophys. Res.*, vol. 97, no. B1, pp. 479–488, 1992.

[14] M. Wyss and R. O. Burford, "Occurrence of a predicted earthquake on the San Andreas fault", *Nature*, vol. 329, no. 6137, pp. 323–325, September 1987.

[15] C. Bufe and D. Varnes, "Predictive modeling of the seismic cycle of the greater San Francisco Bay," *J. Gheophys. Res.*, vol. 98, no. B6, pp. 9871–9883, 1993.

[16] M. Eneva and Y. Ben-Zion, "Application of pattern recognition to earthquake catalogues generated by models of segmented fault systems in three-dimensional elastic solids", *J. Gheophys. Res.*, vol. 102, no. B11, pp. 24513–24528, 1997.

[17] J. M. Carlson, "Time Intervals between characteristic earthquakes and correlations with smaller events: an analysis based on mechanical model of a fault", *J. Gheophys. Res.*, vol. 96, no. B3, pp. 4255–4267, 1991.

[18] M. V. Matthews and P. A. Reasenberg, "Statistical methods for investigating quiescence and other temporal seismicity patterns", *Pure Appl. Geophys.* vol. 126, no. 2–4, pp. 357–372, June 1988.

[19] R. E. Buskirk, C. L. Frohlich and G. V. Latham, "Unusual animal behaviour before earthquakes: a review of possible sensory mechanisms", *Rev. Geophys.* 19, 247–270, 1981.

[20] J. L. Kirschvink, "Earthquake prediction by animals: evolution and sensory perception", *Bull. Seismol. Soc. Am.* 90, 312–323, 2000.

[21] O. Sayeed and S. Benzer, Behavioral-genetics of thermosensation and hygrosensation in drosophila". *Proceedings of the National Academy of Sciences of the United States of America*, 93, 6079–6084, 1996.

[22] S. B. Vanderwall, "Seed-water content and the vulnerability of buried seeds to foraging rodents". *America Midland Naturalist*, 129, 272–281, 1993.

[23] R. Blakemore, Magnetotactic bacteria. *Ann. Rev. Microbiol.*, 36, 217–238, 1982.

[24] J. L. Gould, "The case for magnetic sensitivity in birds and bees (such as it is). *American Scientist*, 68, 256–257, 1980.

[25] J. L. Gould, J. L. Kirschvink and K. S. Deffeys, "Bees have magnetic remanence". *Science*, 201, 1026–1028, 1978.

[26] C. Walcott, J. L. Gould and J. L. Kirschvink, "Pigeons have magnets". *Science*, 205, 1027–1029, 1979.
[27] M. M. Walker et al., "Evidence that fin whales respond to the geomagnetic field during migration", *J. Exptl. Biol.*, 171, 67–78, 1992.
[28] M. M. Walker et al., "A candidate magnetic sense organ in the yellow fin tuna", *Thunnus albacares. Science*, 224, 751–753, 1984.
[29] M. Lindauer, "Recent advances in the orientation and learning of honeybees", *Proceedings XV Int. Congr. Entomol.* pp. 450–460, 1977.
[30] H. Tributsch, "When the snakes awake: animals and earthquake prediction", MIT Press: Cambridge, Mass, 1982.
[31] Maekawa et al., "A statistical study on the effect of earthquakes on the ionosphere, based on sub-ionospheric LF propagation data in Japan", *Ann. Geophys.* 24, 2219–2225, 2006.
[32] Y. Kasahara, et al., "On the statistical correlation between the ionospheric perturbations as detected by sub-ionospheric VLF/LF propagation anomalies and earthquakes", *Nat. Hazards Earth Syst. Sci*, 8, 653–656.
[33] R. A. Grant and T. Halliday, "Predicting the unpredictable; evidence of pre-seismic anticipatory behaviour in the common toad", *Journal of Zoology*, 281, 263–271, January 2010.

9

A GNSS-Based Network

Donatella Dominici, Elisa Rosciano, and Roberta Valerio

Geomatic Laboratory-Afcea — Engineering Faculty-University of L'Aquila

9.1 Aim

This chapter introduces the reader to the concepts related to the permanent GNSS station networks and their use in the study of deformations.

The first part describes what is meant by permanent station and how a definition of PS network was reached. Starting with the global PS networks, mostly used for scientific purposes, regional networks were then created that can be used with the new GNSS methods, such as DGPS and RTK. An overview is given of this development up to the NRTK network in Abruzzo, taken as an example.

Following the earthquake on 6 April 2009, the need was strongly felt to organize a study on the monitoring of deformations and on the analysis of possible seismic precursors, using these permanent GNSS stations.

The concept of deformation is thus outlined and then the chapter focuses on the problem of TEC and of the ionosphere model and the GPS technique. The last part presents the first experimental tests done in the L'Aquila area.

9.2 A GNSS Network

A GNSS geodetic network generally consists of permanent GNSS stations, and its primary function is to provide a materialization of the earth reference system, monitored over time, determined with the greatest precision obtainable.

A PS service network sets the main goal of supplying data, derived products and user support for GNSS surveying, so as to make it easier and more convenient to use (Manzino, 2003). It also has the task of geodetic monitoring and may be useful in the development of further scientific applications, such as the estimating of water vapor in the atmosphere (Dousa, 2001). Briefly, a service network is composed of a set of PS's and a Control and Data Analysis Center that must carry out: the periodic compensation (post-processing) of the network; the estimating in real time of disturbances and errors in PS codes and phases; the modeling of disturbance and error networks and estimating of corrections; the automatic adjustment of the calculation process to changing situations.

Starting in the 1980s, the first networks of permanent GNSS stations were created on a national or continental scale, for prevalently **scientific** purposes (definition of global data, study of movements of the earth's crust, ...). More recently, networks have been established on a **regional scale**, with stations spaced only a few dozen kilometers away from each other. These networks make it possible to provide various types of **positioning services**, with a remarkable improvement in performance and productivity, exploiting the potentialities of differential positioning and RTK.

Differential GPS positioning (*DGPS*): in the area covered by a PS network, the user can work with **just one receiver**, acquiring the differential correction (code) from the nearest station in the network. The acquisition can take place via GSM telephone connection or via Internet by connecting with a Master (web address that transmits corrections 24 hours daily) and using the NTRIP protocol (the necessary software can be downloaded by the user for free).

RTK relative positioning and NRTK: in the area covered by a permanent stations network, the user can work with **just one receiver**, acquiring the RTK correction (phase) from the nearest station in the network. As with the DGPS, the acquisition can take place via GSM telephone connection or via Internet by connecting with a Master and using the NTRIP protocol. The real advantages of the network structure are obtained, however, with **NRTK** (*Network RTK*) techniques, which use the data from several stations simultaneously processed in real time by software installed in the control center, according to different approaches: **VRS** (*Virtual Reference Station*) — the NRTK software generates a correction personalized for the user, calculated according to the position of the Rover, which emulates a station (thus said to be virtual) placed near the user, e.g., at a distance of 1 kilometer from the survey area. **FKP** (*Flachen Korrektur Parameters*) — the NRTK software generates a correction personalized for the user, calculated according to the position of the Rover, with mean correction parameters calculated by interpolation between the surrounding stations in the network. There are other NRTK methods as well, including the **MAX** (*Master-Auxiliary Concept*). Experimenting has shown that all of them give substantially equal results, provided that there is an efficient and, above all, continuous telecommunications connection between the stations and the control center, and between the control center and the user.

Static relative positioning: in the area covered by a PS network, the user can work with **just one receiver**, occupying with it only the points to be determined, with obvious economic benefits. With the positions of the permanent stations known and downloading the **raw data** files acquired from the permanent stations surrounding the survey area, the user is able to calculate the baselines that connect the network stations to the points surveyed, compensate the study network and estimate the coordinates of the points surveyed to an accuracy of a few millimeters. Some permanent station operators also offer, by request, the service of the automatic or manual calculation of the positions of the points occupied by the user.

9.3 Permanent Station

Every permanent GNSS station is equipped with a receiver and a geodetic antenna, and continuously receives the signals (code and phase) emitted by visible satellites, 24 hours a day 7 days a week, and transmits them to a network control center from which they are made accessible to the users.

The choice of installation sites is particularly important in the general plan of a permanent GNSS station network. For each receiver, a site must be found that is free of obstacles and interference and logistically convenient (protected area, accessibility, availability of power and data transmission lines).

The main specifications of the receiver are:

- data sampling at 1 Hz at least;
- RJ45 port that supports Ethernet protocols;
- additional 3 serial interfaces or USBs for further connections; for example, a data backup connection, a connection via local PC and a weather station;
- internal memory in the receiver that can store at least 120 hours of data acquired with 1 second data sampling;

— possibility of real time transmission to the Control Center of code and phase data in proprietary format and in the most recent RTCM format;
— ability to memorize simultaneously the data in the receiver internal memory and to transmit them to the Control Center (independently: FTP push, or controlled by the center).

The following are also considered additional and useful specifications:

— possibility of sampling data at a frequency higher than 1 Hz (5 or 10 Hz);
— possibility of providing a time synchronization signal output;
— possibility of receiving an external synchronization signal (atomic clock) and availability of the relative input port;
— SW for managing pressure, temperature and humidity measurements acquired by a possible weather station connected to the GNSS receiver.

The PS antenna is generally a Choke Ring type with radon protection or equivalent; the coaxial cable between the antenna and receiver is a minimum attenuation type; for sites where the distance between antenna and receiver is greater than 30 meters, it is best to use signal amplifiers along the cable. A UPS must be connected between the electric power supply and a first power port of the PS to guarantee full operation of the PS for at least 120 hours. All the components of the PS must be able to automatically restart following a power failure, with the same configuration they had before the forced shutdown.

The majority of stations are on buildings (57%), but thanks above all to the recent installation of numerous geodynamics stations, there are also a considerable number of monumented stations on the ground, many of which having foundations anchored to the bedrock.

The possible types of monumentation are: reinforced concrete pillar; stainless steel pillar; anchored metal tripod; pylon.

The management and analysis SW for the PS network has the following functions:

— checking the operation of the PS and generating of any alarms;
— adjusting of receiver configuration parameters;

— data transfer and storage;
— management of overall network data;
— data processing (for generating differential corrections and other products).

The checks and operations regarding the individual PS must be possible from both the Control Center and from the station itself.

The SW automatically calculates some parameters for evaluating the proper functioning of the receivers and of the network as a whole.

The Control Center SW must do the following operations:

(1) periodic compensation (post-processing) of the network;
(2) estimating in real time of disturbances and errors in PS code and phase observations;
(3) modeling of disturbance and error networks and estimating of corrections;
(4) automatic adjustment of the calculation process to changing situations.

Every station is equipped with a local acquisition system guided by appropriate software, which thus makes it possible to acquire all the information coming from satellites (codes, phases, ephemerides...).

For each PS it is possible to configure the generation and automatic cancelling of data files (.DAT, SSF, RINEX) at definable hourly intervals, or to show windows on the status of station activities (satellite status, messages).

The data, organized in daily 24-hour files with a sampling step of 30 seconds, are then stored in specific directories both in binary (.DAT) and ASCII format, compressed with dedicated procedures (e.g., Hatanaka).

The daily files for each station are subjected to a Quality Check by developed software and they end up constituting the station's so-called time series. The software, which uses a linear combination of the frequencies carrying the signal, monitors the number of observations acquired, the number of cycle slips, the RMS value associated with the multipath calculated with the C/A and P codes on the L1 carrier (P1 multipath) and with the P code on the L2 carrier (P2 multipath). Indications are also given on the tropospheric delay, on the clock slip, on the elevation and azimuth of the satellites used. The data are collected in a "summary" file. However, any source of unknown noise can be interpreted as multipath, whether of internal origin (instrumental: the noise generated by the receiver electronics) or external origin (electromagnetic interference produced by transmitting antennas that operate at frequencies near GPS frequencies).

The PS's are generally analyzed with scientific software such as Gamit/Globk (developed at MIT, USA), Gipsy (developed at JPL, USA) and the Bernese (developed by the Astronomical Institute of the University of Bern), following specific calculation strategies.

The data processing requires not only the observation files of the individual stations, but also other files that are distributed through the network, at different cadences, from international databanks that carry out this task, such as the *International GPS Service* (IGS). In particular, the daily files necessary are those containing:

- the precise ephemerides of the satellites, available with a delay of about 2 weeks, whereas those provided with a different cadence regard the motion of the pole;
- any problems with the satellites.

The data (24-hour RINEX files, sampled at 30 seconds) are processed daily, producing for each day a coordinates file (.crd), a file containing the covariance matrix of the coordinates (.cov) and a file containing the tropospheric delays (.trp) estimated every hour (of interest for any meteorological applications).

For a reliable evaluation of the data it is necessary to have a time series of at least two years.

The PS's make it possible to obtain greater precision in coordinates and velocity compared to nonpermanent stations, so that in the time span of about three years of data first estimates of velocity can be provided that are useful for the observation and modeling of crustal deformation processes.

In a GNSS positioning service the materialization of the reference system and its distribution among users takes place through the supplying of the estimated coordinates of the permanent stations that make up the system, and thus it is necessary to define an adequate reference system.

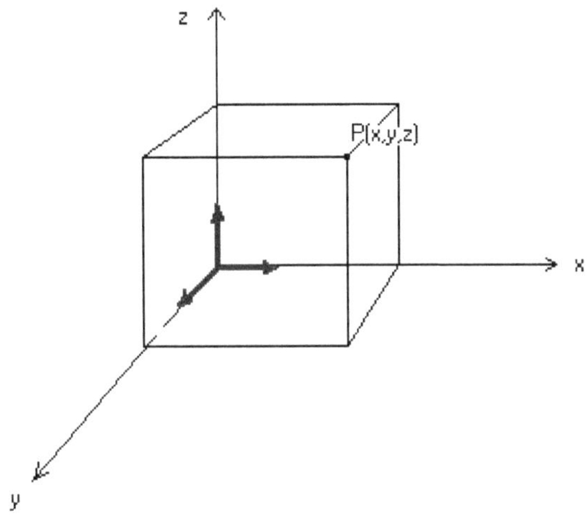

Fig. 9.1 Reference system.

9.4 Reference System

A reference system is defined by means of three unit vectors e_1, e_2, e_3, with a common origin, orthogonal and such as to form a dextrorotatory triad:

$$e_{i*}e_j = \delta_{ij}, \quad e_1 \times e_2 = e_3$$

The coordinates of a point P are represented by the lengths of its orthogonal projections on three axes (Figure 9.1).

In geodesy, the creation of a reference system consists of a catalog of coordinates estimated for a set, called a network, of fundamental points. The users of the reference system utilize these fundamental points and their coordinates as a reference for determining the coordinates of new subnets, said to be "high precision."

Permanent reference systems are created with networks of continuous observation stations (permanent networks), whose coordinates are estimated and continuously monitored, making it possible to guarantee maximum accuracy and consistency.

Static reference systems are created with networks of vertices observed in a single survey from which the coordinates are determined and published just once. Because of this, over the years these coordinates may accumulate deformations on a regional scale, due to differential displacements in the various regions involved.

The global reference systems are created by means of permanent stations all over the globe, the positions of which are estimated using the different satellite geodesy techniques.

9.5 Geodetic Datum

In geodesy, one reference system that makes it possible to express in mathematical terms the position of points on the earth's physical surface or near to it is called Datum.

This is a conventional definition, and in practice it is tied to a series of points materialized on the earth's surface, corresponding to which are given coordinate values.

In geodetic and topographic applications, Earth-Fixed (non-inertial) reference systems are used; in astronomy, systems in which the Earth is in movement (inertial) are considered.

Geodetic datums are differentiated into:

- tridimensional (e.g., WGS84, satellite geodesy);
- planimetric (horizontal datum – e.g., Roma40);
- altimetric (vertical datum, e.g., continental Italy altimetric datum).

In classic geodesy, the definition of datum is based on the concept of *reference surface* and consists in identifying a locally oriented ellipsoid.

A given ellipsoid is conventionally used, the dimensions and shape of which are known, oriented in what is called the point of emanation, establishing that at that point some geometric conditions will take place.

As stated earlier, the definition of datum is tied to a series of points materialized on the earth that constitute a geodetic network, deriving from a group of measurements and the related compensation calculation. The calculation of the network provides the ellipsoidal geographic coordinates of its vertices in the datum used. Therefore the geodetic network defines and materializes the datum by means of the coordinates of its vertices.

The network constitutes the creation of the datum.

Every country has its own geodetic datum; in Italy, due to the historic evolution of the geodetic networks and of cartography, different "classic" definitions of datum are still used today, the main ones including the following:

- National geodetic system, Roma 40: used for national and regional maps ("Gauss-Boaga" plane coordinates), and for cadastral maps in a limited number of provinces;
- ED 50 geodetic system (ED = European Datum): used for the definition of UTM-ED50 plane coordinates, and for the cutting (dividing into sheets) of new and regional IGM maps;
- Cadastral geodetic systems: derived from geodetic systems adopted in IGM works, they are used in cadastral maps and in the GISs using these maps as a base.

9.6 Datum in Satellite Geodesy

In satellite geodesy (currently based mainly on the GPS system), global geodetic datums are used, i.e., valid for the entire world. These are differentiated from those of classic geodesy, which as we have seen are valid locally.

The definition of a global datum is based on a triad of geocentric $OXYZ$ axes, i.e., having the origin coinciding with the Earth's center of mass. The Z axis coincides with the polar axis (Earth's axis of rotation); the X and Y axes lie on the equatorial plane, with the X axis directed according to the prime meridian (Greenwich) and the Y directed so as to complete a dextrorotary triad (Figure 9.2). The geocentric triad is *integral* with the Earth, i.e., it rigidly follows Earth in its motion: thus these systems are called *Earth-Centered-Earth-Fixed.*

In this case also the definition is conventional, considering that the position of the geocenter and the direction of the polar axis (the latter variable in time) are established conventionally. The datum currently used for GPS applications is the **WGS84** (World Geodetic System).

By analogy with classic systems and to facilitate the georeferencing of points (by means of the usual geographic coordinates), the Cartesian triad is associated with a **geocentric ellipsoid**, having a center coinciding with that of the triad and axes oriented according to the XYZ directions (Figure 9.2).

The definition of the WGS 84 datum is actually more complex, and includes a series of physical-mechanical parameters (Earth's mass, rotation speed, etc.).

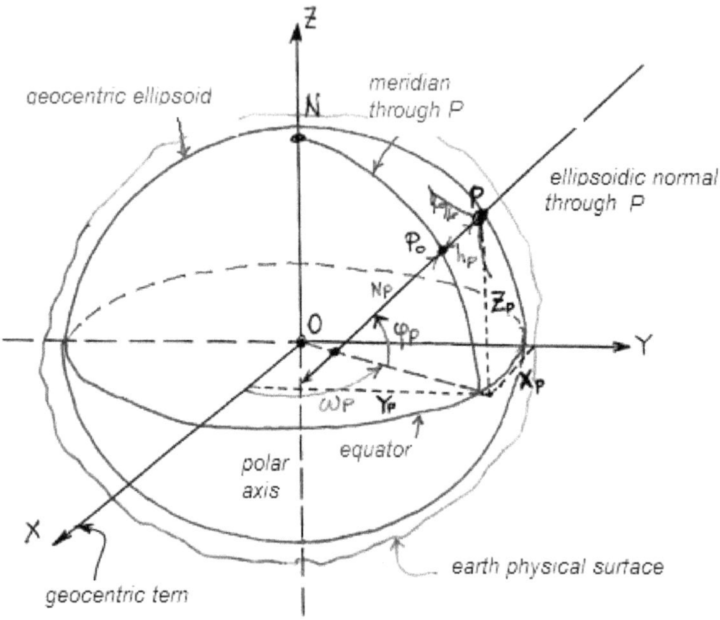

Fig. 9.2 Datum WGS84.

For the purposes of the most common geodetic-topographic applications, however, it is enough to know just the geometric parameters.

The WGS84 system is practically defined and materialized through the institution of various geodetic networks.

In Italy, the geodetic network that creates the WGS84 datum is the IGM95 network, determined by the Istituto Geografico Militare with GPS measurements done in the mid 1990s. This network is a densification of the European EUREF network (which in turn is part of the world IGS network) based on the European ETRS89 datum, integral with European continental shelf and practically coincident with the WGS84.

Then there are local networks created or in the process of being created by various bodies (Regions, cadastre, Provinces, etc.) which further densify the IGM95 network. In recent years numerous permanent GPS/GNSS stations have also gone into operation, with the WGS84 coordinates for many of these being calculated through connections to the IGM95 network. With these progressive densifications, the WGS84 system is materialized and becomes accessible to the user.

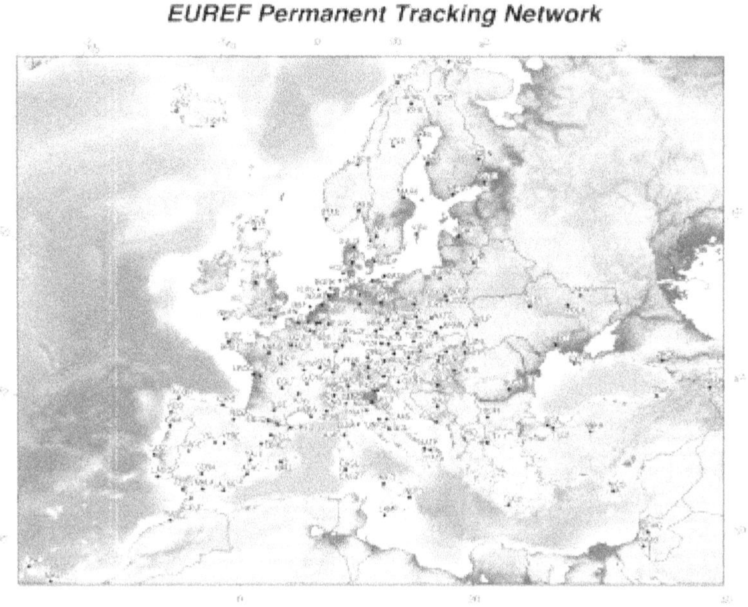

Fig. 9.3 Euref permanent network.

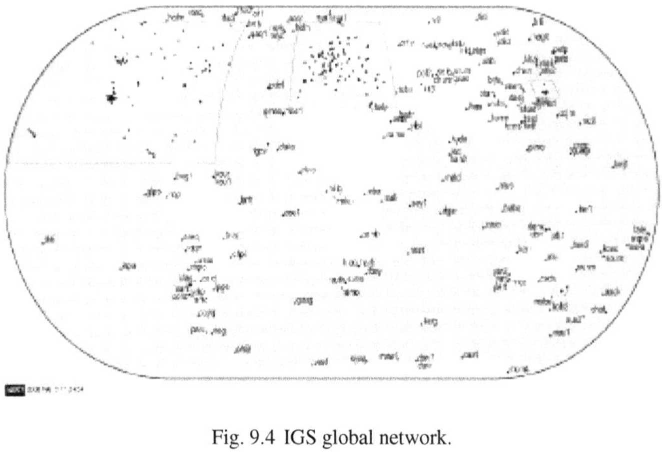

Fig. 9.4 IGS global network.

ICRS — ICRF: International Celestial Reference System and Reference Frame

The XXIII IAU General Assembly in August 1997 established the ICRS (International Celestial Reference System) as the official celestial reference system as of 1 January 1998, replacing the FK5.

Fig. 9.5 IGM95 Italian national network.

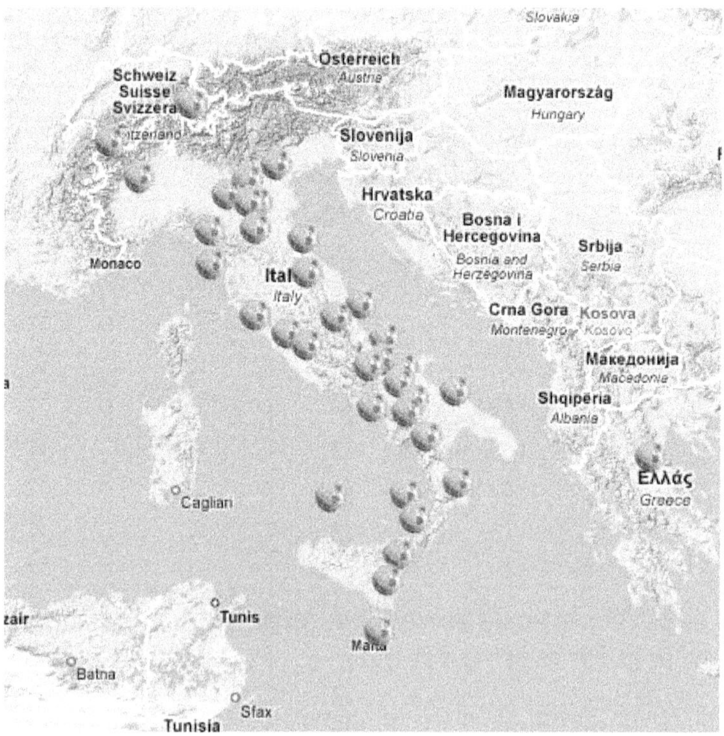

Fig. 9.6 INGV national network.

The physical realization of the ICRS, the International Celestial Reference Frame (ICRF), refers to a set of radio sources (quasars) from outside our galaxy. The realization of the optical frequencies instead consists of the entire Hipporcos catalogue, which includes the FK5 and ICRF positions of a large number of stars. The origin of the ICRS is the center of mass of the solar system, estimated by VLBI observations in conformity with the general theory of relativity, at the conventional date of 1 January 2000 (epoch J2000).

ITRS — ITRF: International Reference System and Reference Frame

With the Celestial Reference System having been defined, the earth reference system must be defined. The basic characteristics of a conventional earth reference system (ERS) are:

- being geocentric: the origin coincides with Earth's center of mass, including the atmosphere and oceans;
- having the Z axis coinciding with the mean position of the rotation axis;
- having the X axis passing through a reference meridian (*Greenwich*);
- having the Y axis such as to complete the dextrorotary triad.

Operatively defined as (ERF) by a set of earth control stations equipped with measurement instruments that take on certain coordinate values, which define the frame with their coordinates.

One example of ITRF is the WGS84, adopted in 1987 by the GPS, a frame defined by the coordinates of over 1500 stations.

The International Earth Rotation Service (IERS) has established and manages an earth reference system called the International Terrestrial Reference System (ITRS) created concretely from the coordinates of a series of stations on Earth that create the International Terrestrial Reference Frame (ITRF) updated every five years. It is based on about 200 stations located on all the tectonic plates so as to be able to monitor continuously the relative movements between the various stations so as to have a more stable definition of the system.

The International Earth Rotation Service was created to:

— monitor the earth's rotation and orientation parameters (EOP);
— create the International Celestial Reference System;
— create the International Terrestrial Reference System;
— monitor the deformations taking place in the earth's crust.

9.7 ITRS Network Realization

The permanent station networks for the realization of the ITRS are essentially those of the three principal methods of geodetic observation:

- Very Long Baseline Interferometry;
- Satellite Laser Ranging;
- Global Navigation Satellite System.

The three networks are created from fundamental points monumented in a stable manner, with observation instruments operating virtually continually over time: permanent networks.

The VLBI global network is coordinated by the International VLBI Service: it consists of about 30 stations that take simultaneous measurements of differences of angles and distances from quasars: the bases between the stations are estimated from the measurements by means of data processing methods.

The SLR network is coordinated by the International Satellite Laser Ranging Service and it consists of about 90 stations, realized by means of adjustable laser guns: a station observes the round-trip time of a laser beam sent by a gun and reflected off artificial satellites orbiting around the planet. The observations allow the modeling of the orbits of the satellites and the estimating of the position of the stations.

The International GNSS Service, IGS, was established in 1993 for the purposes of contributing to the realization and distribution of ITRS, providing GNSS products, defining the standards for permanent GNSS stations and supporting GNSS research. The IGS network consists of about 380 stations; the controlling, monitoring and analysis of the network are handled by various analysis centers, work teams, and pilot projects, as well as a Central Coordination Office.

9.8 EUREF Network: European Reference Frame

The network of permanent EUREF stations was created with the primary goal of creating and maintaining a European ITRF reference system. The latter is the best definition of earth reference system possible today and is materialized under the guidance of the international IERS service, availing itself of GPS and GLONASS satellite positioning. The network now includes about 200 permanent stations spread all across Europe, run by various agencies and with different monumentation, making available daily RINEX files via free access both on http and on FTP with an acquisition frequency of 30 seconds, as well as many other types of information coming from data processing, such as on the data latency for each station and some quality diagrams.

The Agenzia Spaziale Italiana (Italian Space Agency) is one of the 10 bodies that participate in the measurement and calculation of the European continental network of permanent stations.

Every week these centers provide the calculation of various subnets: these data, jointly processed and compensated for long periods, represent the realization of the ETRFaa archives (aa = the year of the calculation). ETRS reference systems originate from this.

The EUREF network is substantially a densification of the IGS network, with the aim of defining the Reference System on a continental scale. It does not provide additional products to the global network.

The network coordinated by the ASI is the natural and necessary medium between the global networks and the service networks, but it cannot guarantee those services possible for more local PS networks; one must go down to the regional level.

9.9 IGS Network

In 1993, wishing to create and maintain a high-precision geodetic reference system on a global scale, the International Association of Geodesy (IAG) commissioned the IGS, which consolidates a world PS network, to calculate and distribute — mainly for scientific applications — the precise ephemerides of GPS and GLONASS satellites, the parameters of the earth's rotation, the corrections of GNSS satellite clocks, the coordinates and velocity of the ITRF stations, ionospheric and tropospheric information and a series of data on satellites, on constellations, on calculations and on the time series of the network and of the subnets calculated from various analysis centers.

In agreement with ITRF, it maintains a global reference system calculated only from GNSS data which is currently called IGS05.

The IGS network now includes over 380 active stations all over the world, run by various agencies and with different monumentation, which make available their raw data free of charge via FTP in RINEX files, with an acquisition frequency of 30 seconds. Today the IGS network (http://igscb.jpl.nasa.gov) is made up of variously located permanent stations: 95 are situated in North America, 23 in South America, 8 in Antarctica, 16 in Africa, 79 in Europe, 31 in Asia and 21 in Oceania.

One very important aspect of the products of the IGS network is connected with the "densification" of the ITRF system on a regional scale. Thanks to the presence of some stations common to different subnets, it is also possible to combine the different solutions with a particular time cadence (monthly, annual or multiannual) to obtain a continental solution at the same time. This is possible through the interaction of that network with the local GNSS networks by using standard SINEX (*Solution Independent Exchange Format*) files that can be read by various data processing programs.

The objectives of the network coordinated by the IGS are:

- the ever more precise definition of a single DATUM (SR) called ITRS, the realizations of which are a set of coordinates called ITRF;
- the evaluation of models of the deformation of the globe (of the tectonic plates).

The coordinates of these vertices are constantly being recalculated; the archives can be found at the Central Bureau of the IGS.

9.10 IGM95 Network

The Geodetic Service of the IGM has the task of creating and maintaining the precision geometric reference for the entire national territory.

The Italian realization of the WGS84 is the GPS IGM95 geodetic network of the Istituto Geografico Militare. The network is complete and uniformly distributed throughout the national territory, and is made up of approximately

2000 vertices having an average density of about one every 20 km, easily accessible and with an average relative precision of 2.5 cm planimetrically and 4 cm altimetrically. The IGM95 network is set up within in the EUREF89 network and is stably materialized.

This network offers users the possibility of doing geo-topographic surveys on the national territory with the GPS and to report the results to the national system. The IGM95 network in fact allows the relation between the WGS84 (ETRS89) and Roma40 systems, providing the coordinates of the vertices in both reference systems and the transformation parameters for each point, with a range of validity of about 10 km. Many Italian regions have densified this network.

Many vertices were connected with the high-precision contour lines and, along with the WGS84 ellipsoidal height, they also provide a precision determination of the orthometric height.

9.11 ASI Network: Geodaf

In Italy, the GNSS networks of permanent stations for positioning services are currently in the stage of realization and development; typically, due to administrative and logistical reasons, at the moment these are independently planned, carried out and managed on a regional scale. The Agenzia Spaziale Italiana (ASI) coordinates many of these stations, some of which are its own.

In over 10 years of activities ASI has organized the collecting of data from 45 parameter GNSS stations, which have been made available to the national and international scientific communities through the GeoDAF databank, located at the Space Geodesy center of Matera (ASI/CGS). The data are distributed via FTP in daily RINEX files Hatanaka compressed according to the standard IGS nomenclature, with data with an acquisition frequency of 30 seconds. In the http site some information is also available regarding the operator of each permanent station and, for some of these stations, the possibility of obtaining the data from the GLONASS constellation.

9.12 INGV GPS Network

The Rete Integrata Nazionale GPS (RING — National Integrated GPS Network) reflects the experience acquired in over 10 years of geodetic activities by the Istituto Nazionale di Geofisica e Vulcanologia (INGV — National Geophysics and Volcanology Institute) which since 2004 has converged into the establishing of a network of permanent stations.

The INGV network was created to conduct geophysical surveys, and for this reason there are problems with the sites chosen connected with the morphology of the area and characteristics of the territory.

At each site there are GPS receivers as well as a series of seismometers and accelerometers connected in real time with three control centers, with the ultimate goal of identifying and monitoring the country's seismic situation.

The network, which is composed of 35 stations spread across the national territory and all run directly by the INGV, makes the data publically available on FTP, with daily files in RINEX format and Hatanaka compressed, with an acquisition frequency of 30 seconds. The network's Internet site also makes available some information on the monumentation for each permanent station, which is generally composed either of a concrete column placed on a building or by a metal tripod anchored by drilling into the ground.

9.13 Abruzzo Permanent Network

The Region of Abruzzo covers an area of 10,794 km^2. Abruzzo's network of Permanent GNSS Stations was created with the aim of providing positioning to all regional users (professionals, governments, etc.), and therefore also users without specific training in processing GPS data, with a precision of under 10 cm and as uniform as possible over the entire regional territory, and with the shortest possible positioning times (for data acquisition).

The Control Center that operates the system of the permanent GPS stations network is entirely unrestricted by their position and it is situated in the seat of the Regional Council in L'Aquila. In its first configuration it considers the situation regarding the stations held to be "priority," i.e., those stations that although remaining fairly distant from each other (but no more than 64 km) may represent an initial possible geometry with a view to serving the most densely populated areas. These stations provide almost uniform coverage of the regional territory, and were positioned attempting to keep the distance between points fairly uniform, from 40 to 70 km, with the aforesaid limits.

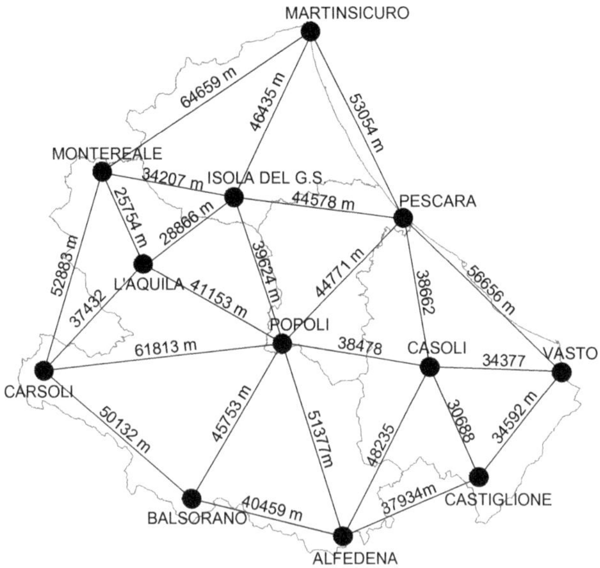

Fig. 9.7 First design of Abruzzo permanent network.

This configuration, in which the average distance between stations is about 45 km, could have been realized considering 12 permanent stations positioned:

on the coast	within the regional territory	on the south — west border
Martinsicuro	Isola del Gran Sasso	Montereale
Pescara	L'Aquila	Carsoli
Vasto	Popoli	Balsorano
	Casoli.	Alfedena
		Castiglione

The configuration of the network of stations considered "priority" is shown in Figure 9.7.

In a second "optimal" configuration, 16 stations were considered; they help to decrease the distance of the longest sides, and if some of them are made more accurately, they may make it possible to check the stability of the reference system materialized by the network of permanent stations.

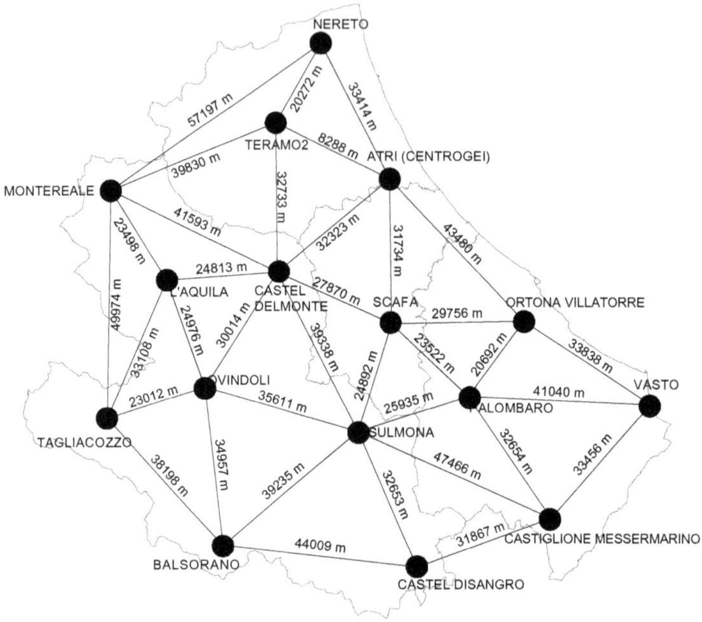

Fig. 9.8 Abruzzo final SP network.

The greatest distance between stations is now 57 km, on the northern border of the region. The average distance instead is about 34 km: this distance can already guarantee the good functioning of the VRS/MRS systems.

As can be seen in Figure 9.8, the sites are partially different from those identified in the "priority" configuration, and for these it is possible to find buildings in the area owned by the Region that can be validly used as PS sites.

The monitoring of any movements in the network can be done by measuring the GPS bases connecting the points of the network and some points (points identified ad hoc within or outside the region) having geologically stable positions and accurate materialization, tied to the PS monumentation site.

The materialization of these points is done with maximum attention given to stability and connection to the ground; therefore structures such as buildings or pillars set on the surface on loose soil are to be excluded.

The possible uses, along with the periodic verification of the infrastructure, are:

— connection to the network of national (ASI) or international (EUREF) PS's;

— geodynamic monitoring of the region;
— subsidence monitoring.

The Region of Abruzzo network can satisfy all technical positioning, with varying degrees of precision (cadastre, roads cadastre, engineering projects, territorial and structural monitoring…), and scientific positioning (maintaining of the Datum, crustal movement studies, water vapor content analysis…).

9.14 Post and Real Time Data Processing

In this case the data acquired by the permanent stations are subject to the necessary quality control to ensure that they satisfy the standards provided; they are generally "thinned out" to epochs of 30 seconds to reduce the size of the files; and they are entered in a databank which can be accessed online by authorized users. Such is the case of the ASI network, whose data are available at the GeoDAF website (http://geodaf.mt.asi.it).

The user accesses the data acquired previously for all operations that can be done in post-processing, essentially for the calculation of the bases in static mode. Indeed, given that in order to limit the size of the data to be memorized the files are generally made available sampled at epochs of 30 seconds, they are unsuitable for rapid, static or kinematic deferred processing, since for this type of surveys the measurements in the field are sampled at a greater frequency.

The permanent stations are able to download data memorized at time intervals programmed by the Control Center, so as to allow access to the data, once the quality analysis has been passed, by users with set and guaranteed delay times, as well as ensuring the achieving of quality standards guaranteed by the system. The Control Center manages the publication of the sites' monographs and the information on the specifications of the receiver and antenna used, and all the metadata necessary for reconstructing any dynamics of the site instrumentation.

This solution does not involve any problems of principle or any particular randomness, although it requires excellent organization and careful maintenance while the service is being carried out.

Real Time Data Processing

In this case the data acquired by the permanent stations are made available to the users with very brief delay times, which depend on the transmission modes, but are so limited as to be able to speak of "real time."

The latency times are in fact sufficiently short as to allow the doing of surveys in real time with the normal geodetic receivers backed only by equipment for receiving the signal sent by the permanent stations; the algorithms in the instruments and the processing software must be able to be the same as those used when there is a master receiver and a rover, both operated by the field operators and with a direct connection between the two receivers.

In order to obtain this, the permanent station must be equipped with the transmitter-receiver necessary for sending the data recorded to the authorized users requesting it at that moment and over the entire area covered by the service; this can be obtained with radio modems, the use of RF radio systems and cellular telephones.

The solution that allows the user to survey in real time with just a receiver and equipment for receiving the signals sent from a station is more expensive and above all more delicate from a functional viewpoint, because conditions outside of the geodetic structure must be guaranteed, basically transmission and reception via modem or cellular telephone.

9.15 Earthquake Preparation Zone

The concept of "Earthquake preparation zone" was developed by various authors many years ago (Dobrovolsky et al., 1979). It is in this area in fact that important local deformations are observed that can probably be traced to the occurrence of future earthquakes. These deformations involve changes in the properties of the crust (density, electrical resistivity, changes in groundwater levels, other geochemical precursors) which can be measured using various techniques. The most important thing to be noted is that all deformations are accompanied by the building up of tension, to the measurement of which GPS techniques have recently made a strong contribution. To determine the size of the earthquake preparation zone some authors use pure seismic precursors like earthquake swarms preceding the destructive event (the magnitudes) and the distribution of deformations, whereas other authors consider the whole complex of physical parameters. There are empirical formulas that correlate the radius of the preparation zone with magnitude or other quantities. Among the phenomena commonly identified as seismic precursors, deformations and variations in the characteristics of the ionosphere may be closely estimated using GNSS techniques.

9.16 Deformations Monitoring

The monitoring of slow deformations of the ground became part of the geophysical methods applied to the study of volcanic and seismogenic areas a few decades ago. In particular, it is possible to obtain the dilatancy measurement parameter from the analysis of the deformation. The so-called "dilatancy theory" seems to explain many of the variations observed before an earthquake as it is closely connected to the behavior of the rocks subject to tension. Indeed, prior to rupturing, countless tiny fractures and gaps form in the rock, with the subsequent increase in volume of the rock due to filling with water coming from the nearby aquifers. This increase in volume appears to be connected with changes in the ground level; the same cause justifies the variations in electrical conductivity of the rocks due to the water that seeps into the gaps and cracks, the higher concentration in well water of radon gas that is released from the fractures and, lastly, the variation in P waves in the area where the earthquake epicenter will be found.

With sufficiently stable points usually taken as permanent GNSS stations, it is possible to obtain the speed of variation of the stations' coordinates, an index of the volumetric variations in the subsoil. Up until now a weighted regression of the planimetric components of the coordinates and from this, through a precise mathematical relationship, the deformation rates of the earth's crust in the form of so-called "strain" tensors were estimated.

To do this, one starts by defining a dataset of coordinates and velocity of displacement of the stations at a given epoch. The repetition of the GPS measurements at the same points during the inter-seismic (time interval between one event and the next) and postseismic stages, makes it possible to define the variations of the position in relation to the points. The job of the analysis of historic series is also to distinguish the study object from all the series of deformations due solely to local or cyclical phenomena that have nothing to do with the crustal deformations begin analyzed. The movements and measurement time are considered linear and the deformation uniform in the zone.

The monitoring done earlier was used to obtain the coordinates and velocity of displacement for each permanent GNSS station, with an average standard deviation of daily positions of 1–2 mm horizontally and about 5–6 mm vertically, and horizontal velocity errors of less than 1 mm/y for time series of 2–3 years. As mentioned, the last result of the analysis will be a two-dimensional tensor of the deformation. The concept of deformation is strictly connected to that of displacement. This can be represented as the variation of coordinates in the direction of the axes compared to the initial coordinates. To clarify this, two station points are considered, the linear extension of the baseline vector between two points of the network (linear expansion or deformation) is expressed as:

$$\varepsilon = \frac{S' - S}{\Delta t \cdot S} \tag{9.1}$$

where:
S' is the length of the baseline vector at the end of the reference time;
S is the length of the original baseline vector

Also, the length of the baseline vector whose azimuth is t is given by:

$$\varepsilon = e_{xx} \cos^2 t + e_{xy} \sin 2t + e_{yy} \sin^2 t \tag{9.2}$$

where:

e_{xx} is the variation in the x direction
e_{yy} is the variation in the y direction
e_{xy} is the shear deformation

For the displacement-deformation connection, if u is the displacement of point A in the figure (a station GPS receiver) and $u + \Delta$ u the displacement of B (another GPS receiver), the deformation will be:

$$\varepsilon = \lim_{\Delta x \to 0} \frac{A'B' - AB}{AB} = \lim_{\Delta x \to 0} \frac{(\Delta x - u) + (u + \Delta u) - \Delta x}{\Delta x}$$
$$= \lim_{\Delta x \to 0} \frac{\Delta u}{\Delta x} = \frac{\partial u}{\partial x}$$

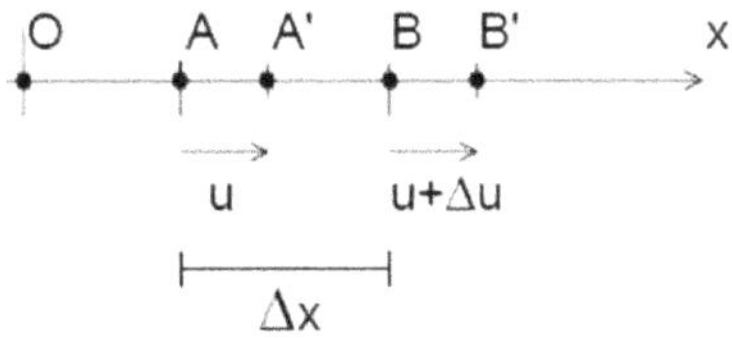

Fig. 9.9 Displacements and deformations.

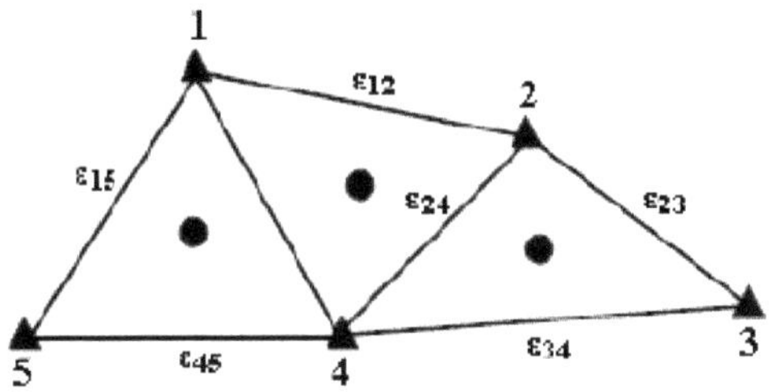

Fig. 9.10 Net triangles sides as strain tensor's components.

The same holds for the component perpendicular to the x direction of the displacement.

The components of the deformation tensor are calculated in the following manner. Consider one of the triangles of the planned network. Its vertices will be the positions of 3 sites whose velocities are known (Figures 9.9 and 9.10).

If the dimensions of the triangle are small compared to the earth's radius, then it can be considered plane. If we suppose that the origin of the system is the triangle's center of mass, once a reference system of axes N and E is defined, the velocities of its vertices are represented by the vector:

$$V = \begin{pmatrix} v_i^e \\ v_i^n \end{pmatrix}_{i=1,2,3} \tag{9.3}$$

Therefore, at the first order, for every vertex i, the equation is:

$$\begin{pmatrix} v_i^e \\ v_i^n \end{pmatrix} = \begin{pmatrix} v_{Bar}^e \\ v_{Bar}^n \end{pmatrix} + \begin{pmatrix} \dfrac{\delta V_e}{\delta e} & \dfrac{\delta V_e}{\delta n} \\ \dfrac{\delta V_n}{\delta e} & \dfrac{\delta V_n}{\delta n} \end{pmatrix} \begin{pmatrix} \Delta E_i \\ \Delta N_i \end{pmatrix} \tag{9.4}$$

This contains 6 unknowns: the 2 components of the center of mass velocity and the 4 components of the gradient of the field of velocities. To solve it requires complex calculations and laborious matrices.

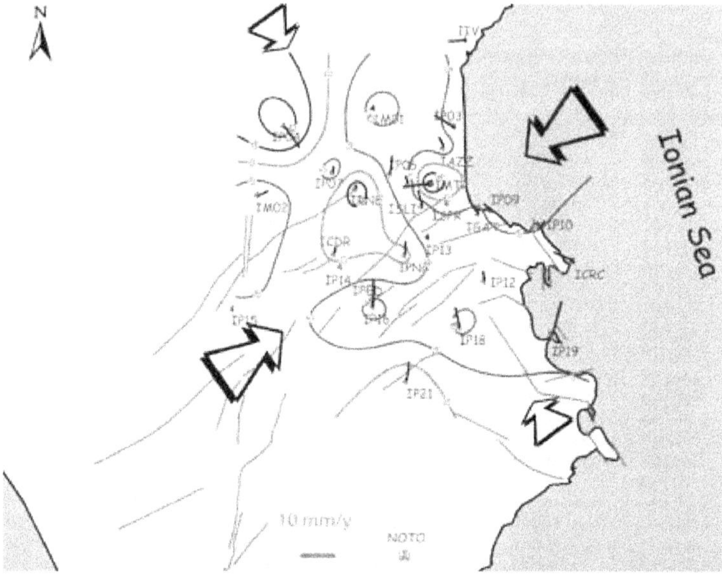

Fig. 9.11 Velocity in GPS stations (blue lines) and vertical deformations (orange and green lines) (mm/year).

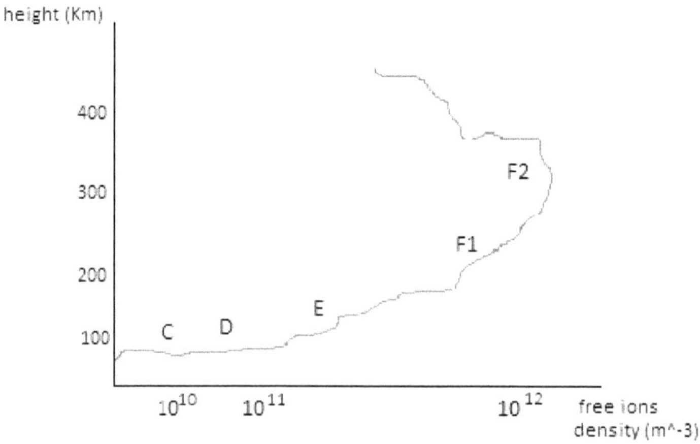

Fig. 9.12 Height — Ions density trend.

The result and the final intent is to interpolate the field of velocities of the vertices to the center of mass of the triangles.

Assuming that the components of this tensor are not just functions of space but also of time, what one wants to know is how much the earth's surface

z, km	N, m^{-3}	n, m^{-3}
200	$4 \cdot 10^{11}$ (day)	$8 \cdot 10^{15}$
300	$2 \cdot 10^{12}$	$5 \cdot 10^{15}$
400	$1{,}5 \cdot 10^{12}$	$3 \cdot 10^{14}$
500	$1 \cdot 10^{12}$	$1 \cdot 10^{14}$
600	$7 \cdot 10^{11}$	$1 \cdot 10^{13}$
800	$(0{,}8 \div 2) \cdot 10^{11}$	$2 \cdot 10^{12}$
1000	$(0{,}4 \div 1) \cdot 10^{11}$	$4 \cdot 10^{11}$
1500	$(2 \div 6) \cdot 10^{10}$	$1 \cdot 10^{10}$
2000	$(1 \div 4) \cdot 10^{10}$	$3 \cdot 10^{9}$

Fig. 9.13 Comparison between electron density N and neutral density in the upper part of the atmosphere.

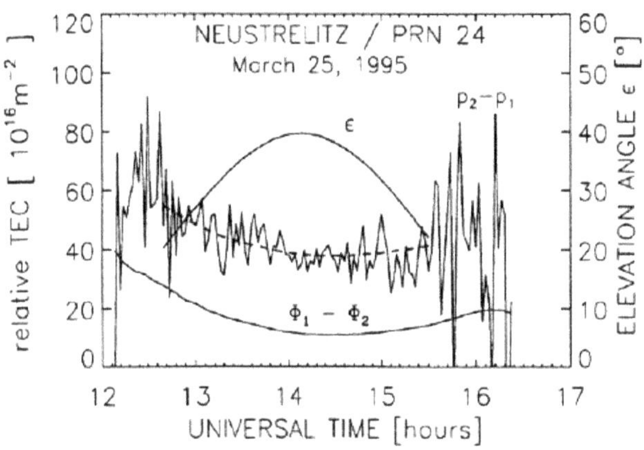

Fig. 9.14 TEC_p's noise trend.

expands or compresses at any given point in the unit of time. In other words, in this way one describes the deformation rate (mathematically it would be the time derivative of the tensor components):

$$\dot{\varepsilon}_{zd} = \begin{pmatrix} \dot{\varepsilon}_{ee} & \dot{\varepsilon}_{en} \\ \dot{\varepsilon}_{ne} & \dot{\varepsilon}_{nn} \end{pmatrix} \tag{9.5}$$

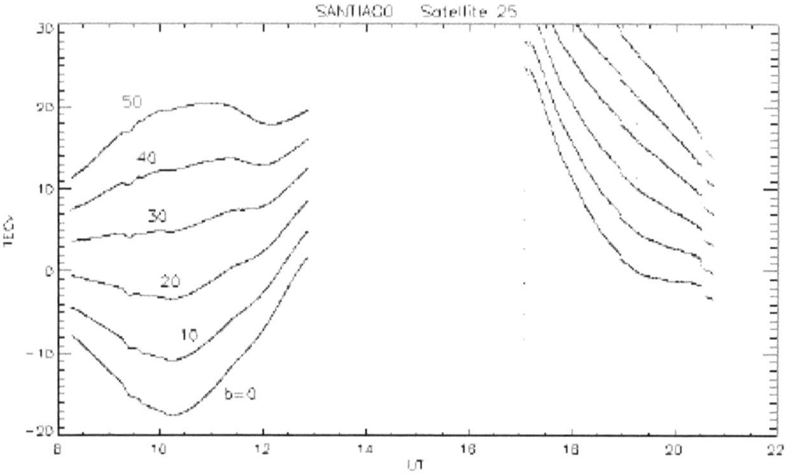

Fig. 9.15 Variation of the TEC_V with b and $I(\theta, \varphi, t)$.

the components of which are related to the partial derivatives of the field of velocity by means of the following equations:

$$\dot{\varepsilon}_{ee} = \frac{\delta V_e}{\delta e}$$

$$\dot{\varepsilon}_{en} = \frac{1}{2}\left(\frac{\delta V_n}{\delta e} + \frac{\delta V_e}{\delta n}\right)$$

$$\dot{\varepsilon}_{nn} = \frac{\delta V_n}{\delta n}$$

from which the matrix can be obtained by integration over time:

$$\begin{pmatrix} \varepsilon_{ee} & \varepsilon_{en} \\ \varepsilon_{ne} & \varepsilon_{nn} \end{pmatrix}$$

After having calculated the tensor components, it is possible to obtain other important parameters:

$$\Delta = e_{ee} + e_{nn} \text{ Dilatancy}$$

$$\gamma_1 = e_{ee} - e_{nn} \text{ Principal shear strain}$$

$$\gamma_1 = 2e_{en} \text{ Engineering shear strain}$$

$$\gamma = \gamma_1 + \gamma_2 \text{ Total shear strain}$$

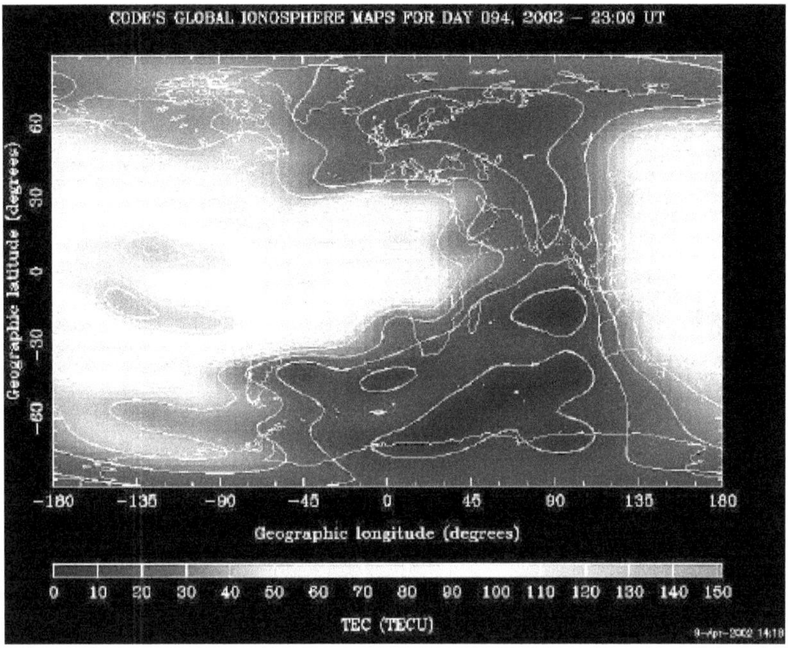

Fig. 9.16 TEC mapping.

From which:

$E_1 = \Delta + \gamma$ Maximum principal shear strain

$E_2 = \Delta - \gamma$ Minimum principal shear strain

$\beta = \arctan\left(\dfrac{e_{\rho e}}{E_1 - e_{en}}\right)$ Direction of maximum principal shear strain arc.

9.17 The Ionosphere

The ionosphere is the highest part of the atmosphere, from 40–50 km up to about 1000 km. It consists essentially of ionized gas resulting from the presence of electrons freed from atoms. For this reason it is also defined and treated as "plasma." A fundamental parameter of this medium is the quantity of free electrons present. The ionosphere in fact exists because of the sun, which causes the atoms in the atmosphere to become ionized. Depending on the density of the electrons, and thus the height, the chemical and physical processes responsible for the formation of the ionosphere change. That is why it is divided into different layers that are described and studied separately (shown in the figure below). They are called C, D, E, F1 and F2. The E, F1 and F2 layers are those of greatest interest, i.e., those in which the greatest density of electrons is found (soft X-rays give up energy to this layer and are thus weakened). The peak concentration of free electrons is always found in layer F2. Layer D is characterized by the presence of γ rays and high energy (hard) X-rays. The C layer is characterized by cosmic rays and the remaining layers by EUV radiation.

As we have just mentioned, the sun regulates the variations in electron density. It varies between day and night and also with the seasons. During the night ionization does not disappear completely, because some time is need for the electrons to bond once again to the corresponding nuclei. If there were no sun for about 10 days, the ionosphere would disappear. Sunspots, characteristic of the sun, act by provoking even more evident variations. However, there is a certain periodicity to this phenomenon, which has been noted to be about 11 years. The closer we move to the earth's surface, the greater the number of whole molecules. As the ionosphere is a conductive medium, it plays a very important role in the global electric circuit that characterizes our planet. Therefore it interacts with the earth's magnetic field, and it is also controlled by it.

The quantitative measurement of the electron density is the Total Electron Content (TEC), expressed in TECU. It is defined as the total number of electrons in the ionosphere contained in an ideal tube having a section of 1 m^2 (*1 TECU* = 1016 electrons/m^2). As the ionosphere shows variations according to the time of day, the season and solar activity, the TEC also shows corresponding variations.

9.18 Ionosphere Model in GPS Technique

Among the errors affecting GPS measurements, biases include those due to tropospheric and ionospheric refraction. In particular, the effect of the ionospheric delay alters the measurement of the satellite-receiver range. The atmospheric bias causes the lengthening of the path in pseudorange measurements and the shortening in phase measurements. The signal anticipation or delay depends on the electron density, variable over time (and difficult to model) in the case of the ionosphere.

Errors depend on the frequency of the signal, thus there will be different errors for the different L1 and L2 carriers (errors of 15–20 m). The error can be modeled (in the case of bases to be measured smaller than 15 km, it is sufficient to use differential or relative methods) by means of the series (Willman-Tucker, 1968):

$$R_{\downarrow}iono = -B_{\downarrow}1/f^{\uparrow}2 - B_{\downarrow}2/f^{\uparrow}4 - \square \tag{9.6}$$

where $R_{\downarrow}iono$ is the ionospheric delay, f is the signal frequency and B_i depends on the electron density, according to

$$B_i = A_i \int_{\rho 0} N^i(h)ds \tag{9.7}$$

with:

N^i = electron density as a function of height h;

A_i = constants to be estimated

ρ_0 = signal path

For high frequencies such as those characteristic of the GPS signal (1575.42 Mhz for the L1 carrier and 1227.60 Mhz for the L2), the second term of equation (1) can be neglected. The final equation will therefore be:

$$R_{\downarrow}iono \cong -B_{\downarrow}1/f^{\uparrow}2\ldots \tag{9.8}$$

Measuring the distance R_i^j between the satellite j and the receiver i with both the L1 and L2 carriers, due to the diversified effect of ionospheric refraction on frequencies f_1 and f_2, two values will be obtained, R_{01} and R_{02} (if $R_{\downarrow}i^{\uparrow}j = R_{\downarrow}01 + R_{\downarrow}iono$):

$$R_{01} = R_i^j - \frac{B_1}{f_1^2}$$

$$R_{02} = R_i^j - \frac{B_1}{f_2^2}$$

from which one obtains, respectively:

$$B_{\downarrow}1 = (R_{\downarrow}i^{\uparrow}j - R_{\downarrow}01)f_1^2$$

$$B_{\downarrow}1 = (R_{\downarrow}i^{\uparrow}j - R_{\downarrow}02)f_2^2$$

Equating the two expressions:

$$B_{\downarrow}1 = (R_{\downarrow}i^{\uparrow}j - R_{\downarrow}01)f_1^2 = B_1(R_i^j - R_{02})f_2^2$$

one obtains the range as a combination of the two frequencies:

$$R_i^j f_2^1 = R_i^j f_2^2 = R_{01} f_1^2 - R_{02} f_2^2$$

$$R_i^j = \frac{R_{01} f_1^2 - R_{02} f_2^2}{f_1^2 - f_2^2} = \frac{R_{01} - R_{02}\left(\frac{f_2}{f_1}\right)^2}{1 - \left(\frac{f_2}{f_1}\right)^2}$$

So the range becomes independent of the ionospheric delay term.

9.19 Ionosphere, TEC and GPS Measurements

The ionospheric effect on observations with dual frequency GPS receivers was used to obtain information on the ionosphere: the difference between the measurements of the two frequencies can be used to calculate the TEC along the signal trajectory between the GPS satellite and the receiver on the ground.

For the code measurement, one finds that the expansion to the first order of the refraction index varies as:

$$n_{ion} = 1 + \frac{c_2}{c^2} = 1 + \frac{C_2 N_e}{f^2}$$

where $C_2 = 40.3\ \mathrm{m}^3/\mathrm{s}_2$, N_e is the electron density and f the frequency.

Thus the correction is:

$$\delta d_{ion} = \int \left(1 + \frac{c_2}{f^2}\right) ds$$

where s is the real path and s_0 is the ideal path in a vacuum. Approximating the first integral along the geometric path then $ds = ds_0$, for which one obtains:

$$\delta d_{ion} = s - s_0 = \int \left(1 + \frac{c_2}{f^2}\right) ds_0$$

which can be rewritten as:

$$\delta d_{ion} = \frac{403}{f^2} \int N_e ds_0$$

Defining the TEC as the total content of electrons along the path of the electromagnetic wave between each satellite and the receiver as $N_e ds_0(N_e$ is the density of the electrons), the delay for the code variable is equal to $\delta d_{ion} = 40.3/\mathrm{f}^2$ TEC. The TEC can be estimated by means of a Taylor series expansion as a function of the latitude and the time; the series coefficients are introduced as unknowns in the equations and estimated by the processing software.

The TEC is also an important geophysical parameter which has applications in the correcting of navigation measurements for single frequency receivers. The TEC, which is a magneto-optical phenomenon, was measured for decades using the Faraday effect (or **Faraday rotation**). Today TEC measurements are done mainly using GPS data (currently there are 379 stations for continuous TEC measurement). Each satellite transmits two different electromagnetic waves, in the L1 and L2 bands. Combining the pseudorange and phase observations,

$$P_i = \rho + c \cdot (dT - dt) + \Delta i_i^{iono} + \Delta i^{\mathrm{trop}} + b_i^{P,s} + b_i^{P,r} + m_1^P + \varepsilon_1^P$$

$$\Phi_i = \lambda_i \cdot \phi_i = \rho + c \cdot (dT - dt) + \lambda_i \cdot N_i - \Delta i_i^{iono} + \Delta i^{\mathrm{trop}} + b_i^{\Phi,s} + b_i^{\Phi,r} + m_1^\Phi + \varepsilon_1^\Phi$$

where:

$i = 1, 2$ corresponding to carrier frequencies L1 and L2
P is the code pseudorange measurement in distance units)
ρ is the geometrical range between satellite and receiver
c is the vacuum light speed
dT, *dt* are the receiver and satellite clock offsets from GPS time
$\Delta iono = 40.3$ TEC/f_i^2 is the ionospheric delay
TEC is the Total Electron Content
$f_i$is the carrier frequency L_i
Δ *trop*is the tropospheric delay
b_i are the receiver and satellite instrumental delays on P and Φ
m_i are the multipath on P and Φ measurements
ε_i are the receiver noise on P and Φ
Φ_i are the carrier phase observations (in distance units)
ϕi are the carrier phase observations (in cycles)
$\lambda = c/f$ is the wavelength
N_i are the unknown L_i integer carrier phase ambiguities

A TEC value from the pseudorange is the following:

$$TEC_\phi = 9.52 \cdot (P_2 - P_1) + instrumental\ delays + multipath + noise$$

However the result is very "noisy"

Therefore, using the phase measurements

$$TEC_p = 9.52 \cdot (\Phi_2 - \Phi_1) - (N_1\lambda_1 - N_2\lambda_2) + instrumental\ delays + multipath + noise$$

which is less noisy than TECp, but more ambiguous.

It was demonstrated that the noise of the TECp grows in correspondence with elevation angles less than 20°, as the figure below shows.

The ambiguity can be removed using the mean between the TECp and TEC ϕ values on a stretch of the satellite's orbit.

$$TEC_L = TEC_\phi - (TEC_\phi - TEC_P)$$

This approximates the TEC resulting in the TECp (characterized by greater disambiguity) but includes instrumental delays, multipath and noise. Nevertheless, it contains the same low noise information as TEC ϕ. Another interesting application is the measurement of the vertical TEC (or local, TECv) which depends only on time and position.

For the detection of TECv a mapping function $M(E)$ is used, with E the elevation angle of the satellite in regard to the receiver. The simplest form of this function is

$$M(E) = \frac{1}{\cos Z}$$

with Z the zenithal angle at the sub-ionospheric point, a point between the satellite and the receiver at the height given by the center of mass of the ionospheric profile, usually between 350 and 450 km (thin shell model). To study the perturbations of the ionosphere, the following formula is sufficient:

$$TECV = TECL * \cos Z$$

However, when the absolute value of the TEC is required, it is necessary to know also the instrumental delays of the satellite and the receiver, as they could be significant.

To obtain these delays and local and global mapping of the ionospheric TEC, it was decided to look for an estimate. The measurement of the TECL between satellite and receiver $T^{rs}(t)$ between receiver r and satellite s at the epoch can be modeled by:

$$T^{rs}(t) = M(E)^* I(\varphi, t) + b^r + b^s$$

where:

$M(E)$ is the mapping function for the elevation E
$I(\theta, \varphi, t)$ is an ionospheric TEC model
θ, φ are latitude and longitude
t is the measurement epoch
b^r, b^s are the differential instrumental delays of the receiver r and satellite s

Given the orbits of the satellite, θ, φ and E are determined from the least squares or from the Kalman filter. For the biases b instead corrections are available online (CDDIS: Crustal Dynamics Data Information System). Simpler methods consist in assuming a priori a value for the TEC of around 3-5 TECU during the night for observations at the zenith.

The variation of the TEC_v with the varying of parameters b of $I(\theta, \varphi, t)$ are given in the figures below.

Lastly, an example of TEC mapping (CODE's Global Ionosphere Map, the latitude and longitude) is given.

9.20 First Examples in the L'Aquila Area

The earthquake of $M = 5.8$ on 6 April 2009 unfortunately caused great damage in the city of L'Aquila and surrounding areas and the death of 300 people. This event was preceded by a preseismic sequence with tremors of magnitudes up to 4 and by an intense postseismic sequence in a southeast direction with regard to the epicenter (Chiarabba et al., 2009).

As written in the preceding paragraphs, the permanent station networks have been planned and realized, with the primary objective of supplying GNSS data so as to make it easier and more convenient to use them (Manzino, 2002) for the geodetic monitoring of the area and also to contribute to numerous scientific applications.

Using the data from these networks and other permanent stations in the Region of Abruzzo and neighboring regions, it was possible to create an important series of spatial data that allowed the study and a first estimate of the "earthquake preparation zone" and the co-seismic surface deformations.

Figure 9.17 shows all the permanent stations taken into consideration for the first analysis of the surface deformations. There are the permanent stations

Fig. 9.17 Permanent stations used for the first results.

of the Region of Abruzzo, 3 RING (INGV) stations, 3 EUREF stations, 1 ASI network station (AQUI), 2 Regional GPS UMBRIA network stations and 2 other stations run by the GEOTOP company.

Daily data were collected with sampling at 30 seconds from 1 February to 2 May 2009. The time distribution range was chosen so as to be able to point out any displacements, but also according to the need to be able to collect and process the data rather quickly.

9.21 Processing Data and First Analysis of the Results

The data collected were processed to obtain a first solution. Data was used from 64 days before the earthquake and 27 days after the earthquake, adopting the international standards in the raw data processing by Bernese 5.0 software. An outlier rejection was done with modeling of the time series to estimate discontinuities. A minimum constraint adjustment was performed, fixing the coordinates of EUREF permanent station of Terni (UNITR) in the EUREF datum. To check the stability of the Reference System, a Helmerth six parameter roto-translation was carried out between the first and last solutions: the check did not show significant variations. Code Precise ephemerides were used.

For the April 6th event, a discontinuity is noted in the time series of the permanent stations, generally greater vertically then horizontally. Some stations then stabilized around new values, whereas that of Paganica and

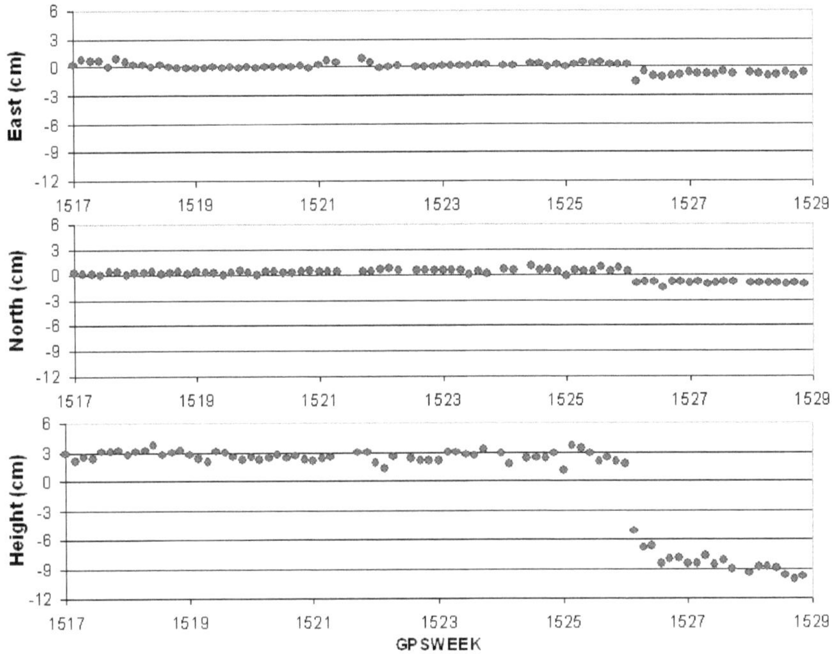

Fig. 9.18 PAGA-Paganica permanent station: pre e post sismic data.

Aquila continued to show movements as a result of seismic swarms after April 6th.

Figure 9.18 shows the Paganica time series. These displacements were obtained considering the difference between the average value of the preseismic and postseismic solutions and excluding the period from 5 to 13 April, in which some stations continued to show vertical displacements. The negative vertical displacement at the Paganica permanent station (PAGA) is evident and shows a lesser component moving towards the west and south.

As seen in the preceding paragraphs, the TEC analysis is an appreciable precursor of earthquakes, but it shows its limits when the magnitude is not very high (5.5). The tremors occurring before and after the main tremors are not detected by this approach. It is well known that there are strong variations in the ionosphere parameters as a function of season and solar cycle, and it is quite possible that ionosphere sensitivity to the seismic events and the overall precursor's characteristics may also change with the solar and season cycle phase. Investigations on this question are possible at this time.

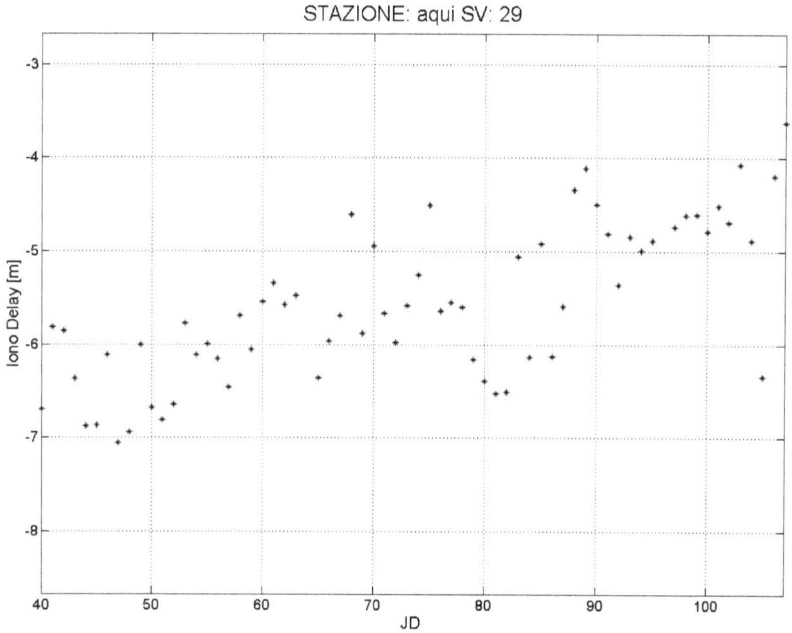

Fig. 9.19 Ionospheric delay in AQUI–SV 29.

Figure 9.19 shows the time series computed for the L'Aquila GPS permanent station (AQUI) in the range of 33 Julian Days before and after the April 6th earthquake, considering Satellite 29. The first test developed and the first analysis demonstrate that the ionospheric delay can be an additional earthquake precursor (De Agostino and Piras M., 2010). These tests can be taken for each permanent station and computed using "one way positioning."

9.22 Conclusions

In the last thirty years the GNSS technique has undergone enormous development and has give a strong boost to new research in the geodetic-topographic field. The time series of data from permanent GNSS stations and the increasingly denser distribution of these stations in the territory make a significant contribution to the study of the pre- and postseismic signal and the study of deformations. The possibility of finding a link between the seismic event and the variation in the ionospheric component of the GNSS signal, which may be correlated to other seismic precursors, is of great interest.

Acknowledgements

Prof. Dominici offers her sincerest thanks to all the colleagues who allowed her to keep working after the April 6th earthquake, which caused the destruction of most of the Faculty of Engineering at the University of L'Aquila. There would not be much of the data extracted in the last section were it not for the work of the group from the Politecnico di Milano (Sansò, Biagi), Politecnico di Torino (De Agostino, Piras) and the Faculty of Engineering at the University of Perugia (Radicioni, Stoppini).

References

[1] L. Baroni, F. Cauli, D. Donatelli, G. Farolfi, and R. Maseroli, "La rete dinamica nazionale (RDN) ed il nuovo sistema di riferimento ETRF2000," *Servizio Geodetico — Istituto geografico Militare — Firenze*; pp. 1–9.

[2] L. Biagi, "I Fondamentali del Gps" — *Workbooks*, vol. 8 2006, pp. 22–35.

[3] L. Biagi, S. Caldera, M. Crespi, A. M. Manzino, A. Mazzoni, M. Roggero, and F. Sanso, "Una rete GNSS di ordine zero per i servizi di posizionamento in Italia: alcune ipotesi e test", *Atti 11° Conferenza Nazionale ASITA, Centro Congressi Lingotto*, Torino 6–9 November 2007, pp. 1–2.

[4] D.Dominici " Tecnica NRTK e la rete di stazioni permanenti GNSS abruzzese" pp. 13–21. Atti dell'Istituto Italiano di Navigazione June–July 2008.

[5] L. Biagi, S. Caldera, D. Dominici, and F. Sansò, "The Abruzzo Earthquake: temporal and spatial analysis of the first geodetic results" EUREF workshop 2009.

[6] Dousa, J., On the specific aspects of precise tropospheric path delay estimation in GPS analysis, IAG Symposia 2001, vol. 125, Springer, 2001.

[7] M. Pesenti, Alberto Cina a Ambrogio Manzino, "Misura e controllo di deformazioni con metodi GPS", DITAG (Dip. di Ingegneria del Territorio dell'Ambiente e delle Geotecnologie), Politecnico di Torino, C.so Duca degli Abruzzi 24-10129, Torino, — (manuele.pesenti, alberto.cina, ambrogio.manzino)@polito.it, pp. 1–2.

[8] Esposito, Alessandra, "Studio della deformazione geodetica delle isoleeolie con particolare riferimento al vulcano di panarea," Tesi di dottorato di ricerca in geofisica, XIX ciclo, Dipartimento di Fisica — Settore di Geofisica, ALMA MATER STUDIORUM UNIVERSITA' DI BOLOGNA, 2007.

[9] Sergey Pulinets and Kirill Boyarchuk, "Ionospheric Precursor of Earthquakes," pp. 1–88, pp. 15–18, Springer-Verlag Berlin Heidelberg, 2004.

[10] V. I. Keilis-Borok, P. N. Shebalin, and I. V. Zaliapin, "Premonitory patterns of seismicity months before a large earthquake: five case histories in southern California." The National Academy of Sciences, 2002.

[11] A. M. Manzino, "Stazioni Permanenti GNSS in Italia: scopi, usi e prospettive." Invited paper_-ASITA, Perugia 2002.

[12] Dominici D., G. Fastellini, and F. Radicioni, "The 6 April L'Aquila earthquake: different approaches for the evaluation of surface displacements." Gi4DM2010 Conference-Geomatics for Crisis Management-Torino.

[13] M. De Agostino and M. Piras, "Earthquake forecasting: a possible solution considering the GPS ionospheric delay." Gi4DM 2010 Conference-Geomatics for Crisis Management-Torino.
[14] I servizi di posizionamento satellitare per l'e-government — Ludovico Biagi,Fernando Sansò Editors- Geomatic Workbooks, vol. 7 WEB SITE References IGS: http://igscb.jpl.nasa.gov/ Rete SP Abruzzo: www.regioneabruzzo.it RETE IGM95: www.igm.it Progetto Italiano Monitoraggio: www.INGV.it.

CURRICULA VITAE

Donatella Dominici

Donatella Dominici is Associate Professor for the area ICAR 06 (Geomatic) at the L'Aquila University, Faculty of Engineering.

She deals with teorical and experimental researchs on GPS technique, in GNSS Abruzzo NRTK planning and in the development of technologies in geodetic networks. In the last years her studies involved on High Resolution Satellite Images and worked in regional projects to defend the coastal Abruzzo area. She is in the COSMOCOAST programme, is in editorial boarding of Applied Geomatic (Springer Verlag) , in Scientific SIFET (Society of Surveying and Photogrammetry) Group.

Roberta Valerio

She is graduated in Environmental Engineering with honors in 2008 in the Faculty of Engineering of L'Aquila. In the same faculty she is attending a PhD in geomatic. Actually her research is on Gis Techniques to plan in reconstruction stages, and on deformation monitoring of damaged historic building in L'Aquila after the 6th April earthquake.

Elisa Rosciano

She is graduated in Environmental Engineering with honors in 2010 in the Faculty of Engineering of L'Aquila. In the same faculty she is attending a PhD in geomatic.

Her research is on GRASS Open source applied on archeological sites and on laser scanning data analysis.

LIST OF ABBREVIATION

ASCII	American Standard Code for Information Interchange
ASI	Italian Spase Agency
CGS	Space Geodesy Center
DGPS	Differential GPS Positioning
ED50	European Datum 1950
EOP	Earrh Orientation Parameters

EUREF	European Reference Frame
FKP	Flachen Korrektur Parameters
FTP	Fyle Transfert Protocol
GIS	Geographical Information System
GLONASS	Global Orbiting Navigation Satellite System
GNSS	Global Navigation Satellite System
GPS	Global Position System
GSM	Global System Mobile
IAG	International Association of Geodesy
IAU	International Astronomic Union
ICRF	International Celestial Refence Frame
ICRS	International Celestial Reference system
IERS	International Earth Rotation Service
IGM	Istituto Geografico Militare
IGS	International GPS Service
INGV	National Geophisic and Volcanology Institute
ITRF	International terrestrial Reference Frame
MAX	Master-Auxiliary Concept
MRS	Multi Reference Station
NRTK	RTK Network
NTRIP	Networked Transport of RTCM via Internet Protocol
PS	Permanent Station
RF	Radio Frequency
RINEX	Receveir Indipendent Exchange
RING	National Integrated GPS Network
RMS	Root Mean Square
RTK	Real Time Kinematic
SINEX	Solution Independent Exchange Format
SLR	Satellite Laser Ranging
TEC	Total Electron Content
UPS	United Parcel Service
USB	Universal Serial Bus
UTM	Universal Transverse Mercator
VLBI	Very Long Baeline Interferometry
VRS	Virtual Refence Station
WGS84	World Geodetic System 1984

10

System for Predicting Natural Disasters

Claudia Corinna Benedetti*, Massimo Buscema†
and Marina Ruggieri‡

**ONPS (Permanent National Observatory for Security), Italy*
†*Semeion — Research Center of Sciences of Communication, Rome, Italy*
‡*University of Roma Tor Vergata — Center For TeleInFrastructures (CTIF_Italy)*

10.1 Aim

The Chapter brings to the conclusions of the book highlighting risk scenarios and proposed solutions to cope with them.

10.2 Awareness of the Level of Exposure to Risk

When in Italy a disastrous event occurs that can be traced back, at least in its main determining factors, to natural causes, we "rediscover" the innate fragility of our national territory.

Italy, as an irrefutable fact, is a country where the exposure to the risk of natural disasters (tectonic earthquakes, hydrological alterations, vulcanic eruptions, etc.) is particularly elevated, though not reaching the levels that characterize some "critical areas" sadly famous at the international level, such as Japan and some areas of Mexico and the United States.

Regarding these issues, there perhaps perdures a process of collective removal, with the exception of that quick, dramatic gaining of consciousness of the problem on the occasion of events that, due to their gravity, call the attention of the entire nation to the difficulties of the populations struck and the consequent economic and enviromental damage.

A more deeply-rooted and diffused awareness of the level of exposure to risk would have allowed, over time, for the development of an adequate policy of prediction aimed at reducing the territory's vulnerability. We must consider that if the risk of certain disastrous events appears somewhat unavoidable, the same cannot be said for the quantity of damages incurred.

10.3 Conscious Behavior and Good Practice

The defense of the environment and the climate should represent for our generation and future ones, a categorical imperative, above all when we consider how slowly the awareness of the need to change matures in our consciousness. This awareness is of a need to change our various types of behavior and the lifestyles for attitudes and actions that will allow us to contribute, concretely and daily, to foreseeing disastrous events.

To reach such an objective — certainly a challenge — one must appeal to each citizen's sense of responsibility, beginning with young people and with the schools. It is necessary to teach them also, and above all, how one can defend oneself in the case of natural events such as, for example, earthquakes and floods. Ignorance can take more victims than the events themselves.

The abuses in construction are another cause: constructing buildings without geological norms in a given territory; building thousands of houses around

Vesuvius; cementing around waterways to build villas on a river or a lake, or under a hill with the risk of landslides (cf., for example, Sarno).

We must appeal to a sense of responsibility that, through adequate information and formation, may allow people to comprehend the absolute necessity of putting into act, immediately, a series of behaviors and good practices to favor a sense of well-being and security, of sharing in the choices that make everyone feel an integral part of the future of our country.

It is a new educative model towards which we must tend, made of co-responsibility but also of concrete examples (such as that of the teacher from San Giuliano in Puglia who, during the earthquake of 2002 made her students get under their desks because she in turn had learned it during a course; a small gesture that saved them from the tragedy that instead struck 27 children and their teacher) and of tests, for instance, of evacuation.

Natural disasters are, in fact, affirming themselves as the primary problem of human cultures: the more such cultures grow in quantity and quality, the more earthquakes, landslides, volcanic eruptions and other disastrous events seem unbearable, both for life and for the economy. Yet our planet has always shaken, experienced landslides, and flooded.

Diffusing a new culture also means not accepting passively the events that put our lives and those of our loved ones at risk.

Usually the communities react to these events with politics, with technology and with science:

a. They get organized and inform the populations at risk;
b. They build human settlements in a more intelligent manner;
c. They attempt to predict catastrophic events in advance.

But to be able to move quickly in this direction, politics needs greater courage, while science and technology need a great imagination, challenging objectives, and great rigor.

Politics must have the courage to understand that catastrophic events, called natural disasters, can today be something different than an unpredictable whim of planet Earth. Nature, in fact, has a logic to the way it functions, that often comes into constrast with our interpretation of events.

To confront such phenomena in the right manner, the political world must assume that prevention is not only possible, it is a duty and indispensable.

It is not by chance that the United Nations, with the support of the European Union, evaluating the enormous costs, in terms of human lives and environmental human lives and damages, which are faced yearly on a global disasters, has recommended the development of initiatives decisively oriented towards predicting them.

In macroeconomic terms Italy, among the countries of the OECD, is that which makes the largest financial contributions, and should therefore be among the countries most interested in setting up a similar systemic approach.

After some disastrous events, such as the earthquakes of Friuli and Irpinia, just to offer an example, the Italian scientific community organized itself immediately to diffuse knowledge of the risks and to increase consciousness of having to live with them; but this is not enough.

It is necessary to face, in a systematic manner, the problem of reducing the geological risks by taking advantage of the theories and methods that science puts at our diposal and favoring the development of innovative technologies.

10.4 Brief Excursus Regarding A Few Past Earthquakes that Have Occurred in Italy

To begin this brief historical excursus, historians retain that it is possible that the quake registered in Rome in 849 had its epicenter in Abruzzo.

Only 4 of the 13 earthquakes of a magnitude superior to 5.5 that have occurred in Italy from 1961 until today have been preceded by a seismic sequence with at least one quake of a magnitude of 3.8 or greater, one month before the event and within a radius of 50 kilometers from the epicenter: Belice 1968; Friuli 15 September 1976; Umbria-Marche 1997; Aquila 2009; at San Giuliano in Puglia there were also foreshocks inferior to 3.9.

The INGV has chosen this intensity as decisive. In the other 9 cases there were no shocks of $M > 3.8$ that preceded the larger quake.

How many dozens of shocks have occurred in Italy with an equal or superior intensity without having given rise to a strong earthquake?

The quantity of earthquakes registered in Abruzzo is considerable and there are also discrepancies among sources. For example, there are those who cite the earthquake of 27 November 1456 as the most terrible in historical times for Central Italy: the event is not mentioned in the INGV's parametric catalogue of Italian earthquakes.

Giuseppe Mercalli mentions a strong shock on 20 September 1731; the catalogue lists one on October 15.

In 1315 it seems that only Aquila was hit. A few years later, in 1349, the shock is framed within the ambit of a strong seismic crisis throughout all of Italy that caused damages all the way to Rome.

Even 27 November 1461 would seem to be an isolated event and one of local interest. It is to be noted that there were many repeats of this shock, and a very strong shock 20 days later. The INGV (an Italian National Institute of Geophysics and Volcanology) catalogue reports an earthquake in 1498.

In 1703 (almost 250 years of silence, a period in which no destructive earthquakes occurred), there was one of the strongest earthquakes of the Central Appenines: despite the fact that the epicenter was probably close to Norcia, Aquila suffered considerable damage.

The shocks once more had effects even in Rome and were followed days later by a rather strong aftershock.

Then there is the earthquake that devastated southern-central Italy, of 1730/32. Giuseppe Mercalli writes that in September 1731 "numerous earthquakes occurred in Abruzzo, including an especially ruinous one on the 20th"; the INGV catalogue reports an event on October 15 two days before the earthquake that shook Foggia.

Mercalli's notes appear rather precise and coincide with numerous dates reported in a book by Giovanni Flores entitled "*Il Terremoto*" (*The Earthquake*).

Seismic Risk Map in Italy

10.5 Mitigation of Seismic Risk

An effective strategy for mitigating seismic risk requires an answer to some questions:

(a) where, when and how strong can an earthquake strike the Region in question?

(b) what consequences should be expected if such an event were to happen?

The answer to the first question regards predicting earthquakes, while the second is the object of studies on seismic risk.

Predicting earthquakes consists in identifying the magnitude, location, and time of origin of a future seismic event, with such precision as to allow a single assessment of the success or failure of the forecast itself.

The prediction can include an intrinsic percentage of false alarms and also of failure to predict. Nevertheless, the predictive capacity of the method considered must be superior to that which can be obtained by declaring alarms in a casual manner, or on the basis of characteristics of seismic activity independent of the time.

The precision with which one can predict the spatio-temporal localization of a strong earthquake, that is, of an event with a magnitude superior to a certain threshold, M°, is still an open problem.

The spatial uncertainty in locating the epicentre of an impending earthquake is intrinsic and cannot be inferior to the dimensions of the source of the earthquake. A seismic source is, in fact, an object of finite dimensions, physically representable as a portion of the fault immersed in the lithosphere. In addition, it is necessary to consider that the precursors can manifest themselves in an area much more extensive than the source itself.

Predicting earthquakes, just like the effects induced by them (in terms of the shaking of the ground), can be effected either by a probabilistic approach or by utilizing the available knowledge regarding the physical processes that are the basis of the generation of seismic events and of the propagation of the waves.

However, the characterization of the statistical properties of the phenomenon in question requires the use of a relevant number of observations. These are generally not available above all for what regards the strongest, and thus less frequent, earthquakes.

The difficulty in assessing the occurrence of earthquakes and the propagation of their effects makes it desirable to apply methodologies that allow the use of a larger body of geophysical and seismological knowledge and new data able to furnish more realistic and physically consistent indications.

10.6 Considerations of the "Second International Seminar on Prediction of Earthquakes" — Lisbon, 29–30 April 2009

After twenty years since the First International Seminar on Prediction of Earthquakes held in Lisbon in 1988, organized under the auspices of the U.N. And the European Community (formerly ECE), the results obtained thanks to new lines of research and new technologies for the observation and analysis of earthquakes were discussed in Lisbon in the aforementioned Seminar, the "Second International Seminar on Prediction of Earthquakes".

The Seminar allowed scientists to define the state of the art of the different methodologies and observations connected to the prediction of earthquakes. The program, with the coordination of international experts, covered a wide spectrum of disciplines, from geochemistry to paleo-seismology, from seismological studies to electromagnetic, geodetic, and multi-parametric studies.

Concluding the Seminar was a discussion and approval of a "resolution" whose main points are summarized as follows:

— *"Despite the progress made in the last twenty years, the prediction of earthquakes remains a complex problem, whose solution requires advanced and long term research and observations*;
— *It is opportune to promote a strategy of international multi-disciplinary collaboration, to guarantee the independent evaluations granted by the advanced technological resources available. In particular the data, both historical data as well as that instrumental data that is updated in timely fashion, should be made available to the scientific community, in such a way as to allow*: (a) *an estimate of their quality*, (b) *the application of new methods of research and* (c) *the comparison, over the course of time, between different experiments.*
— *Given the progress made in the acquisition, elaboration and modelling of data, as well as in the multi-disciplinary*

interpretation of the precursory phenomena, it is advisable to have a more intense co-operation and exchange of information between experts of the different disciplines.

It is desirable that Institutional Entities and Non-Governmental Organizations interested in the prediction of earthquakes define test areas, consensually and making use of the support of diverse independent experts, adequately financing the activities of research in this field.

— *It is necessary that any information available in real time, regarding the prediction of an earthquake, shall be communicated in a timely fashion to the competent authorities, according to rules that give guidance on the accuracy and reliability of forecasting methodologies.*
— *Finally, it is desirable to have an appropriate activity of education and formation.*"

It concludes by affirming that "*Therefore the efforts currently focused on the very costly activity of relief and recovery should rather be directed towards preventative actions, which are more convenient and effective*".

10.7 The Precursors of Earthquakes

The precursors of earthquakes, observable on the earth's surface or in close proximity to it, must regard quantifiable and statistically significant phenomena.

According to what was established by the sub-commission on the Prediction of Earthquakes, instituted by the International Association of Seismology and Physics of the Earth's Interior (Wyss, 1997 and included references), the criteria to establish the significance of a precursory phenomenon (that is, an anomaly that precedes a strong earthquake) can be synthesized in this way: the anomaly must be:

(1) due to the mechanisms that lead to earthquakes;
(2) simultaneously detected in more than one site or by more than one instrument;
(3) the anomaly, and its relation with the subsequent occurrence of the earthquake, that is, the rules according to which prediction is to be carried out, must be precisely defined;

(4) both the anomaly and the rules must be taken from a body of data independent of all the data used for the prediction of earthquakes.

Among the many signals that have been proposed as precursors useful for predicting earthquakes, the following are indicated:

— abnormal variations in seismicity;
— variations in the speed and the spectral characteristics of the seismic waves and of the source mechanisms;
— crustal deformations on a regional scale;
— abnormal changes in crustal strains;
— variations of the geomagnetic and gravitational field, telluric currents and resistivity (geoelectric precursors);
— abnormal modifications of the groundwater flow and the content of diverse chemical components in the water (Rn, F, CO2, Nitrogen Oxides);
— anomalies in atmospheric pressure, temperature and flow of terrestrial heat.

Until now the efficacy of the majority of the phenomena proposed as precursors has proved inadequate or, at most, not demonstrated, above all due to the absence of systematic and sufficiently prolonged and observations.

Strong earthquakes, in fact, are rare events and each phenomenon considered a precursor is characterized by its own fluctuations, not linked to seismicity, that make it particularly difficult to detect the precursory signal.

Such an obstacle is overcome, at least in part, when one considers the seismic precursors identifiable in the catalogues of earthquakes, which contain instrumental observations that are systematic, prolonged, and commonly available, and which therefore allow for a verification, on a vast scale, of the seismic anomalies proposed as precursors of a strong earthquake.

The results of the most recent studies regarding the identification of possible precursors, presented in the course of the Second International Seminar on Prediction of Earthquakes, confirm the difficulty of separating the anomalies caused by the strong impending earthquake from variations that have a different origin (e.g. Raininess, micro-seismicity), variations of the temperature and/or of the atmospheric pressure, etc.).

This is evident from the inter-disciplinary project conducted along a segment of the North-Anatolic fault, in North-West Turkey, starting in 1985 (Woith et al., 2009, Volume of Abstracts cited above), that offers the analysis of nine different parameters (including measurements of the deformation of the surface, of the field of gravitational and geomagnetic field, of the level and temperature of the subterranean waters, of radon emissions, as well as the monitoring of micro-seismicity).

Therefore, the most promising approach seems thus to be represented by the predictions based on specific changes of background seismicity within a defined area.

10.8 The Prediction of Earthquakes in Italy

In Italy numerous studies have been dedicated to the precursors of earthquakes, to identify the characteristics of premonitory seismic activity of the last centuries. A few seismic studies have been carried out but practically no experiment for validating such precursors through real predictions has been carried out until now.

For instance, the attempt to indicate the zones in Italy where it is most probable for strong earthquakes to occur, through probabilistic analyses carried out within areas of smaller dimensions, cannot be evaluated without defining *a priori* the threshold of probability that determines the area alerted. Furthermore, the statistical procedures utilized seem very little linked due to the small number of data available for most of the Regions in question.

10.9 Some Data (1)

6 million people live in high hydro-geological risk zones

The lack of information regarding the most appropriate use of land, the abandonment of old hydraulic-agrarian and hydraulic-forestal systemizations, especially in our region, the eccessive anthropization of the territory, the use of inappropriate farming techniques, the aggravating phenomenon of forest fires (cf. *incendi boschivi*), have led to a notable intensification of the erosive processes that have become known, above all in recent years, due to the striking consequences they have caused the territory with the loss of human lives and economic damages.

— 3 million live in zones of high seismic risk.
— 22 million citizens inhabit zones of medium risk.
— 10% of the territory and 89% of Italian townships are afflicted by elevated hydro-geological criticality.
— Almost 50% of the whole national territory and 38% of townships are at elevated seismic risk.
— 1 million 260 thousand buildings are at risk of landslides and flooding; of these, more than 6 thousand are schools, 531 are hospitals.
— 19% of the at-risk population, that is, over a million persons, lives in Campania.
— 825 thousand in Emilia Romagna; more than half a million in each one of the large northern Regions, Piedmont, Lombardy and Veneto. It is in these Regions, together with Tuscany, where persons and things are most exposed to dangers, due to the elevated density of settlements and the size of territorics that record situations of risk."

Data on seismic risk; 40% of the citizens live in zones with an elevated risk of earthquakes; the buildings that are used mainly for residences were built before the anti-seismic law for construction.

The Regions with the greatest surface area at elevated seismic risk_

— Sicily with 22.874 sq.km and almost 1 and half million buildings, of which roughly 5,000 are schools and 400 are hospitals; Calabria with 15 thousand sq.km and more than 7 thousand buildings, including 3,130 schools and 189 hospitals and Tuscany with almost 14,500 sq.km, more than 560 thousand buildings of which almost 3 thousand are schools and 250 are hospitals.

(1) Data taken from the study *Terra e sviluppo, decalogo della terra 2010 — Rapporto sullo stato del territorio italiano*', (Land and Development, a Decalogue of the Earth 2010 — Report on the status of the Italian territory') done by the *Centro Studi del Consiglio Nazionale dei Geologi* with the help of Cresme.

Elevated seismic risk regards almost almost 50% of the entire national territory and 38% of the townships. It is estimated that along the surface of high seismic risk (50% of the Italian territory) there have been built roughly 6 million 300 thousand buildings, of which 28 thousand are schools and 2,188 are hospitals. In the balance are also 1 million 200 thousand buildings, for residential and non-residential use, that have been built along the 29,500 sq.km of territory with an elevated risk of landslides, mudslides, floods; of these more than 6 thousand are schools and 531 are hospitals.

What emerges even more pointedly is that the themes of the ordinary maintenance of the territory, of risk prevention, of local and national responsibility in the choice of the placement of buildings, of the need for quality territorial planning, and of making the territory secure are the true public works that need to be done in our country, and urgently.

The millions of Euros allotted until now only serve to buffer the disaster, to repair the damage. It is necessary to make Italy secure with a plan of prevention, able to link the citizens' safety with the relaunching of the local economies, and that may also combat abusive constructions and reckless urbanization.

10.10 The Legislation in Italy

Under the legislative profile, there are some lacunae. In fact, the law 138/96 does not provide for any institutional organ to oversee civil projects. The body responsible for this verification, the Civil Engineering Department, has been almost stripped by the norm itself of its many powers of supervision. It controls only 10% of the projects presented (selected by lottery), and does not address the question of a project's merit but checks only the presence of the written projects and the technicians' acceptance of responsibility for what they affirm.

In fact, they do not control the calculations nor the geological report. They do not control whether the plot has been investigated with surveys. They do not verify the work of the constructor, except through inspections often done when the structures have already been completed.

It would therefore be desirable that the Civil Engineers Department be given back its original powers of supervision and sanction to ensure greater safety of buildings.

Often, to save money, "metric penetrate tests" are conducted. These tests are considered completely useless and inadequate by the same scientific community. Generally, in fact, there is spent in geognostic studies roughly a fifth of what should be spent. This also occurs for public structures, which should be even safer than civil ones.

The "New Consolidated Text on Constructions", infringed for years, "New Technical Norms on Construction–2008), is aimed at safeguarding human safety. It dictates all the rules for planning in an anti-seismic manner, obligatory for strategic structures. However, it does not provide for punishments for transgressors.

Furthermore, the new norm should, as in many other civilized countries in Europe, require to carry out geo-technical studies (coring surveys, samplings, laboratory tests, tests on site and in the laboratory, studies of local seismic response) before the carrying out of every structure that will interact with human life. This concept has been well received by many European states, but not yet by our own, despite Italy being considered the "birthplace of law." Yet Italy, particularly central Italy, is one of the most seismic areas in Europe.

10.11 Our Project for Prevention

Prevention with regard to earthquakes and possible eruptions, as well as to hydro-geological risk has never been taken into consideration by any government. There now exists an unavoidable need to make the government able, with a strongly innovative and scientific project, to preserve our country from the risks it faces.

Prevention means acting before something that we do not like happens. Regarding engineering, we may thus recognize:

a. **A constructive prevention**: to ensure that the construction of human settlements takes place from the outset in safer locations, with materials and technologies able to withstand highly aggressive natural events, and with information technologies capable of alerting and organizing the competent institutions quickly.
b. **A proactive prevention**: aimed at coming up with scientific systems capable of best predicting the probability of where, when and how catastrophic events will occur, in sufficient time before the event. It is evident that if the scientific effort in this direction is not increased, the damages to our country will be incalculable.

   If we pause to direct our attention to the past, in the '40s, meteorological predictions seemed impossible to many; today they are a part of our activity of social and economic planning. If we think about the second world war, it was that terrible event that allowed us to transform meteorology into a science.

   Science also exists to make what seems impossible today, probable tomorrow and a reality the day after.
c. **A retroactive prevention**: aimed at planning systems of immediate intervention, for catastrophes that have occurred or are in progress, and to be organized in human and technological terms to try to limit damages as much as possible and prevent further damage. This is the only sector in which, until now, our country has demonstrated great organizational capacity and efficiency.

Instead, constructive prevention as well as proactive prevention are still in a neonatal phase. The growth of the first depends in great part on the political will to convince and oblige the various social actors to utilize a technology that is already available and/or easily adaptable.

Proactive prevention is a question of study and scientific experimentation, but it is that which, when it works, costs the least and tends to level also the human and economic costs of the catastrophes. However, it is the most difficult to accomplish.

Predicting an event means being familiar with its dynamic, at least approximately, in time and/or for a certain period of time.

10.12 The Prerequisites of Prevention

The complexity of certain processes is such that the majority of people consider them unpredictable. Chance is invoked. But in the course of human history, chance has often proved to be a metaphor for our lack of foresight and sometimes ignorance.

The first prerequisite for knowing and predicting the dynamic of a process consists in the humble work of gathering data in the time and space of that process and its environment, its surroundings. This is only the first step, indispensable but not sufficient.

The second prerequisite consists in equipping oneself with a mathematical, modelistic and experimental system, for the purpose of creating hypotheses. These would then be translated into experiments by means of computer and blind verifications would be run. But even this is not enough.

The third prerequisite regards the new knowledge that recent scientific activity has expressed regarding the nature of complex systems.

A complex system is a system that does not follow fixed rules, but generates, changes and adapts them to itself in the course of its lifespan.

A human being is a complex system; an antique watch or a modern computer are simply complicated systems.

10.13 Artificial Adaptive Systems

The processes that nature generates, catastrophic and non, are complex systems. Therefore, they cannot be known through a mechanistic science, able to repair watches and build computers.

To comprehend the dynamic of a complex system it is necessary to have a mathematics that does not impose its rules of functioning on the data of the system in question. Rather, almost like a midwife, it allows the global behaviour of the system to emerge from the local interaction of its components, data and equations.

These new mathematical algorithms are named Artificial Adaptive Systems. Their purpose consists in causing the global logic of the system to emerge from the data, which represents significant parts of the system itself. Like in nature, from the bottom up, from the dialogue between cells to the human being.

Life itself goes from the simple to the complex; the opposite direction is the path of death.

No logic of death will make clearer to us the path of life.

The fourth prerequisite once more regards the knowledge that we have recently acquired on complex systems.

A complex system is not the sum of its parts, just as a human being is not a well assembled set of organs.

This means that when we try to comprehend a natural process there is no guarantee that the data we need are only those that are closest, in terms of time and space, to the process itself.

Often a simple toothache has its causes in the position of the foot when one walks.

It is probable that a natural process, such as seismic activity, has its reasons not 200 km from the site of the catastrophic event, or from the nearest fault, but depends upon a complex interplay of constructive and destructive waves that regard all the seismic activity of planet Earth, the geomagnetic activity of the Sun and Jupiter, the lunar phases and certain aspects of the climate.

In practice, to predict a local phenomenon it is very likely that we need data that is global, in time and space.

Our planet, in fact, is not simply a well assembled globe mounted in a glass case, isolated from its context.

10.14 The Importance of the Use of Local and Global Information in Our I-SWARM Project

To predict complex natural processes, including natural catastrophes, we need mathematical models able to process information ranging from the local to the global and, at the same time, global data that allows for the prediction of local (global) phenomena by activating Proactive Prevention of natural disasters: to predict them before they happen.

In the light of the preceding considerations in function of a scientific interest that might have repercussions in the industrial realm, a work group has been created composed of some partners with specific and diversified competencies in the sector: Permanent Observatory on National Security (ONPS), Space Software Italia S.p.A. (SSI), the University of Rome Tor Vergata-CTIF Italy, the University of Pisa, Semeion Research Center, Expert Systems, ITP elettronica.

The work group has elaborated the I-SWARM project, which aims at improving Proactive Prevention in our territory, that is, to come up with scientific systems capable of best predicting the probability of where, when and how catastrophic events occur, in a reasonable amount of time before the event.

For the first time data could be collected together, and in an incremental manner, regarding the signals of natural processes from earth, from the sky and from open sources (sensations of the people from the site, abnormal animal behaviours):

1. Local data:
    a. From land: Swarms of sensors on samples of buildings in order to signal the micro-movements and twists;
    b. From the sky: GPS and SAR data to monitor the zones of interest;
2. Global data:
    c. From land: Local and global seismic and geological data;
    d. From the sky: Local and global meteorological data;
3. Local and global data from open sources:
    e. Data gathered in special call centres for reports of unusual occurrences, verified and codified in a special database using a semantic search engine.

For the first time new Artificial Adaptive Systems will be used, already tested in the medical field and in the field of defence, to merge all the available databases in results for predicting various types of natural disasters: earthquakes, mudslides, landslides, collapses, floods, etc.

An independent scientific committee will oversee the various phases of the project.

Rigorous validation protocols will be provided in a blinded fashion to assess the predictive capacity of the various applications.

The level of knowledge of the nature of the seismic phenomena, the capacity to gather, transmit and elaborate enormous quantities of data and signals, but above all the capacity to consider data coming from multiple sources with the possibility of merging them through intelligent models of elaboration, characterize the I-SWARM project as a strongly innovative project.

10.15 Conclusions

The project's goal is to make it possible to predict seismic activity sufficiently in advance to enable the authorities to put into action operative interventions both for the population as well as in the areas in which to focus their safety interventions.

The project will begin by applying the models of elaboration to data regarding tectonic and volcano-tectonic earthquakes, to set the premise for its successive extension to other types of disastrous events.

The aim is to activate in Italy a Centre for Research on the Prediction of natural disasters that in a few years would be able to furnish predictions that are useful to the competent institutions.

The project, therefore, is not meant to substitute existing structures, but is meant to involve the efforts that the entire national community must face to defend itself actively from the inevitable risks underlying the territory, in an organized vision, thus making a prudent management of the environment one of the cornerstones of a development capable of generating new resources and new knowledge.

Ever looking to the past, to grasp proactive ideas: if the Navy had not challenged the sea, today we would still be travelling by canoe.

Index

RIVER PUBLISHERS SERIES IN COMMUNICATIONS

Other books in this series:

Volume 1
4G Mobile & Wireless Communications Technologies
Sofoklis Kyriazakos, Ioannis Soldatos, and George Karetsos
September 2008
ISBN: 978-87-92329-02-8

Volume 2
Advances in Broadband Communication and Networks
Johnson I. Agbinya, Oya Sevimli, Sara All, Selvakennedy Selvadurai, Adel Al-Jumaily, Yonghui Li, and Sam Reisenfeld
October 2008
ISBN: 978-87-92329-00-4

Volume 3
Aerospace Technologies and Applications for Dual Use A New World of Defense and Commercial in 21st Century Security
General Pietro Finocchio, Ramjee Prasad, and Marina Ruggieri
November 2008
ISBN: 978-87-92329-04-2

Volume 4
Ultra Wideband Demystified Technologies, Applications, and System Design Considerations
Sunil Jogi and Manoj Choudhary
January 2009
ISBN: 978-87-92329-14-1

Volume 5
Single- and Multi-Carrier MIMO Transmission for Broadband Wireless Systems
Ramjee Prasad, Muhammad Imadur Rahman, Suvra Sekhar Das, and Nicola Marchetti
April 2009
ISBN: 978-87-92329-06-6

Volume 6
Principles of Communications: A First Course in Communications
Kwang-Cheng Chen
June 2009
ISBN: 978-87-92329-10-3

Volume 7
Link Adaptation for Relay-Based Cellular Networks
Bas¸ak Can
November 2009
ISBN: 978-87-92329-30-1

Volume 8
Planning and Optimisation of 3G and 4G Wireless Networks
J. I. Agbinya
January 2010
ISBN: 978-87-92329-24-0

Volume 9
Towards Green ICT
Ramjee Prasad, Shingo Ohmori, and Dina Simuni´c
June 2010
ISBN: 978-87-92329-38-7

Volume 10
Adaptive PHY-MAC Design for Broadband Wireless Systems
Ramjee Prasad, Suvra Sekhar Das, and Muhammad Imadur Rahman
August 2010
ISBN: 978-87-92329-08-0

Volume 11

Multihop Mobile Wireless Networks

Kannan Govindan, Deepthi Chander, Bhushan Jagyasi, Shabbir N. Merchant, and Uday B. Desai

February 2011

ISBN: 978-87-92329-44-8

For Product Safety Concerns and Information please contact our EU representative GPSR@taylorandfrancis.com
Taylor & Francis Verlag GmbH, Kaufingerstraße 24, 80331 München, Germany

www.ingramcontent.com/pod-product-compliance
Ingram Content Group UK Ltd.
Pitfield, Milton Keynes, MK11 3LW, UK
UKHW022220190926
475702UK00003B/468
* 9 7 8 8 7 9 2 3 2 9 5 7 8 *